ISBN 978-0-266-96155-0
PIBN 10916068

CURTIS'S
BOTANICAL MAGAZINE;
OR,
Flower-Garden Diſplayed:

IN WHICH

The moſt Ornamental FOREIGN PLANTS, cultivated in the
Open Ground, the Green-Houſe, and the Stove, are
accurately repreſented in their natural Colours.

TO WHICH ARE ADDED,

Their Names, Claſs, Order, Generic and Specific Characters, according
to the celebrated LINNÆUS; their Places of Growth,
and Times of Flowering:

TOGETHER WITH

THE MOST APPROVED METHODS OF CULTURE.

A   WORK

Intended for the Uſe of ſuch LADIES, GENTLEMEN, and GARDENERS, as
wiſh to become ſcientifically acquainted with the Plants they cultivate.

CONTINUED BY

JOHN SIMS, M. D.
FELLOW OF THE LINNEAN SOCIETY.

VOL. XXIII.

Que votre eclat eſt peu durable,
Charmantes FLEURS, honneur de nos jardins!
Souvent un jour commence et finit vos deſtins,
Et le ſort le plus favorable
Ne vous laiſſe briller que deux ou trois matins.
Ah! conſolez vous en, Jonquilles, Tubéreuſes,
Vous vivez peu de jours, mais vous vivez heureuſes.
M. DE DESHOULIERES.

LONDON:

Printed by STEPHEN COUCHMAN, Throgmorton-Street.
Publiſhed at No. 3, St. GEORGE'S-CRESCENT, Black-Friars-Road;
And Sold by the principal Bookſellers in Great-Britain and Ireland.
M DCCC VI.

CORRIGENDA ET ADDENDA.

No. 661, *pag. alt. l.* 24, 25, for " in having feffile germs" read " in not having feffile germens."

No. 709, *pag. alt. l.* 10, dele " are."

No. 916, *pag. alt. l.* 5, for " their" read " the."

No. 912, *l.* 1 & 9, for " CAMPANULA MACROPHYLLA " read " C. ALLI-
" ARIÆFOLIA," and add to the fynonyms

CAMPANULA *alliariæfolia. Willd. Sp. Pl. Salifbury in Parad. Lond. t.* 25.

CAMPANULA orientalis Alliariæ folio, flore albo pyramidato. *Tourn. Cor.* 8.

By a ftrange overfight we neglected to obferve, that this fpecies of Campanula had been already taken up by WILLDENOW and recently defcribed by our friend Mr. SALISBURY, and figured in the Paradifus Londinenfis, on which account the name of *alliariæfolia* ought certainly to have been retained.

INDEX.

In which the Latin Names of the Plants contained in the *Twenty-Third Volume* are alphabetically arranged.

INDEX.

In which the Englifh Names of the Plants contained in the *Twenty-Third Volume* are alphabetically arranged.

Printed by S. Couchman, Throgmorton-Street, London.

Syd Edwards del   Pub by T Curtis, St Geo Crescent Nov 1. 1805.   F landon sculp

PROTEA STELLARIS. STARRY PROTEA.

✱✱✱✱✱✱✱✱✱✱✱✱✱✱✱✱✱✱✱✱

Clafs and Order.

TETRANDRIA MONOGYNIA.

Generic Charaɛter.—Vid. N⁼⁼. 878.

Specific Charaɛter and Synonyms.

PROTEA *ſtellaris ;* foliis ovato-lanceolatis carnoſis apice
calloſis, capitulo hemiſphærico glabro terminali
involucrum radiatum concolor ſubæquante.

DESCR. A low ſhrub. *Stem* very much branched from the
baſe ; branches ſome ſimple, others verticillately divided at the
upper part. *Leaves* ſeſſile, crowded, without order, lanceolate, narrowed towards the baſe, fleſhy, concave, terminated
with a ſmall callous point, thoſe on the upper part of the
flowering branches ſmooth, but ſome of the lower ones and
thoſe on the younger ſhoots hairy underneath. *Flowers* yel-
lowiſh green in a hemiſphærical terminal head, not downy,
ſurrounded at the baſe with a regular, radiated, ſmooth *involucre,* projeɛting ſomewhat beyond the head of flowers, of the
ſame colour and ſhape as the leaves. Within the involucre
are two or three rows of boat-ſhaped *braɛtes,* hairy underneath, ſhorter than the tube of the corolla, but the upper part
of the receptacle is without paleæ. *Corolla* four-petaled : petals
linear, revolute, villous without, ſlightly adhering downwards
into a tube. *Style* ereɛt, exſerted : ſtigma club-ſhaped, ſmooth,
greeniſh.

    This plant is very nearly allied to PROTEA *pallens* and
conifera, two ſpecies, the varieties of which, THUNBERG allows
to be very difficultly diſtinguiſhed from each other ; nor is it
certain that the plants this Botaniſt has deſcribed under theſe
names are the ſame as thoſe of LINNÆUS.

We

We fhould not have hefitated to confider our plant as one of the varieties of *conifera*, and of the particular one figured by Breynius, were it not for·the total want of all woollinefs in the head of flowers. At the fame time the fhortnefs and greater regularity of the involucre which has the fame colour as the leaves, and the verticillate branches, feparate it from *pallens*, with which the fmoothnefs of the head unites it. In this dilemma we have thought it fafeft to confider this plant as an intermediate fpecies, diftinct from both.

A native of the Cape of Good Hope, of as eafy culture as any of the genus.

Our drawing was taken from a plant communicated by Meffrs. Napier and Chandler in June laft; we were foon after favoured with another by our friend Mr. Samuel Curtis of Walworth.

●

S.J. Edwards del. Pub. by T. Curtis S. Geo Crescent Nov 1. 1805.

HIBISCUS PALUSTRIS. MARSH HIBISCUS.

✷✷✷✷✷✷✷✷✷✷✷✷✷✷✷✷✷✷

Clafs and Order.

MONADELPHIA POLYANDRIA.

Generic Charaɛter.

Cal. duplex : exterior polyphyllus. *Capf.* 5-locularis polyfperma.

Specific Charaɛer and Synonyms.

HIBISCUS *paluftris ;* caule herbaceo fimpliciffimo, foliis
ovatis fubtrilobis fubtus tomentofis, floribus
axillaribus. *Sp. Pl.* 976. *Willd.* 3. *p.* 808. *Hort.*
Kew. 2. *p.* 454. *Mart. Mill. Diɛ. n.* 2. *Cavan.*
Diff. 3. *p.* 162. *t.* 65. *f.* 2. *Fabric. Helmft. n.* 18.
ALTHÆA paluftris. *Baub. Pin.* 316.
ALTHÆA hortenfis feu peregrina. *Dod. Pempt.* 655.
ALTHÆA Sida diɛa quibufdam. *Baub. Hift.* 2. 957, (quoad
defcriptionem, fed figura eft alterius plantæ.)
KETMIA paluftris flore purpureo. *Tourn. Inft.* 100.
HIBISCUS *Mofcheutos ;* foliis ovalibus, acuminatis, dentatis,
fubtus incano-tomentofis; nonnullis fubtricufpi-
datis : pedunculis *quafi* petiolis innatis : corolla
ampliffima : capfula extus glabra. *Michaux Flor.*
Bor. Am. 2. *p.* 47. ?

The external calyx confifts of twelve linear leaflets, the internal is five-cleft : fegments ovate, acute, quite entire. The
corolla has apparently five petals, but they cohere at the bafe.
Stigmas remarkably large and globofe. Capfule fmooth without.
Seeds globular, fhining. The peduncles are folitary, axillary,
jointed near the extremity, the length of the petioles, with
which they adhere at the bafe. The leaves vary, fome of them
being fimply oval-acuminate, others fomewhat three-lobed
owing to the elongation of the lateral nerves.

The

The Marſh Hibiſcus is a native of North-America, and ſeems to have been very early introduced into Europe, being mentioned by Dodonæus as a foreign plant cultivated in the gardens of Holland. It does not occur under this name in Michaux's Flora of North-America, but we ſuſpeƈt that what he has deſcribed, as the Hibiscus *moſcheutos* of Linnæus, is the ſame plant; indeed we very much doubt whether both ſpecies are not in reality the ſame. Be this as it may, we have no doubt but that our plant is the *paluſtris* of Linnæus, having had an opportunity of comparing it with a ſpecimen from Kalm in the Bankſian Herbarium, which however does not appear to differ from the Hibiscus *Moſcheutos* of the ſame colleƈtion. In both, the peduncle and petiole appear rather to be united at the baſe than to grow the one out of the other.

Is a perfeƈtly hardy herbaceous plant, but will rarely flower in our gardens without the aid of artificial heat.

Syd. Edwards del   Pub by T Curtis St Geo Crescent Nov 1 1805   F. Sansom sculp

Cor. 4- feu 5-petala, calyci infidens. *Cal.* 1-phyllus, ventricofus. *Capf.* tricocca.

Specific Character and Synonyms.

EUPHORBIA *petiolaris;* petiolis verticillatis folio orbiculato longioribus, floribus folitariis, caule fruticofo inermi nodofo.

For this hitherto undefcribed fpecies of Euphorbia we are indebted to the Marquis of BLANDFORD, who obligingly communicated the fpecimen from which our drawing was taken, from his Lordfhip's colleƈtion at White Knights in Berkfhire, in May laft. It is nearly allied to EUPHORBIA *cotinifolia,* from which it is chiefly diftinguifhed by its flenderer and more woody ftem, by the petioles, inftead of being fimply oppofite, growing four or more in a whirl, and being longer in proportion to the fmall leaves, and by the flowers being folitary, whereas in *cotinifolia* they grow in a dichotomous panicle.

We find a fpecimen of the fame fpecies in the Bankfian Herbarium under the name which we have adopted, colleƈted by Mr. RYAN, from whence we learn that it is a native of the Weft-India Iflands.

Should be kept in the dry ftove with the other tropical fpecies of Euphorbia.

And. F. Edwards del   Pub by T Curtis, St Geo Crescent Nov 1 1805   F Sansom sc

Aster Reflexus. Reflexed-Leaved Star-Wort.

****

Claſs and Order.

SYNGENESIA POLYGAMIA SUPERFLUA.

Generic Charaƈer.

Recept. nudum. *Pappus* ſimplex. *Cor.* radii plures 10. *Cal.* imbricati ſquamæ inferiores patulæ.

Specific Charaƈer and Synonyms.

ASTER *reflexus;* fruticoſus, foliis ovatis ſubimbricatis re-
curvatis ſerrato-ciliatis, floribus terminalibus. *Sp.
Pl.* 1225. *Reich.* 3. *p.* 803. *Willd.* 3. *p.* 2015.
Mart. Mill. Diƈ. n. 2. *Amæn. Acad.* 6. *Afr.* 68.
Berg. Cap. 285.
ASTER *reflexus. Bot. Repoſ.* 93.
ASTER africanus fruteſcens ſplendentibus parvis et reflexis
foliis. *Comm. Hort.* 2. *p.* 55. *t.* 28. *Raii Sup.* 159.

A flower ſo nearly reſembling the common Daiſy, would hardly attraƈ the attention of any, did not its unuſual foliage afford ſome appearance of novelty. The only other recommendation it poſſeſſes is its continuing to bloſſom through the winter. The ray of the flower is ſaid by MARTYN to be blood-red, a tranſlation of LINNÆUS's *radius ſanguineus;* but the colour is confined to the tips and on the under ſide only, the upper ſide of the ray being perfeƈly white.

Native of the Cape and a hardy greenhouſe ſhrub, eaſily propagated by cuttings, or by ſeeds, which it ſometimes produces with us.

Introduced, according to Mr. DONN, in 1790.

S.Edwards del.   Pub. by W.Curtis, S.t Geo. Crescent Nov.r 1 1805   Sansom sculp.

VERBASCUM PHŒNICEUM. PURPLE-FLOWERED MULLEIN.

✸✸✸✸✸✸✸✸✸✸✸✸✸✸

Clafs and Order.

PENTANDRIA MONOGYNIA.

Generic Charaĉter.

Cor. rotata; fubinæqualis. *Capf.* 2-locularis, 2-valvis.

Specific Charaĉter and Synonyms.

VERBASCUM *phœniceum;* foliis ovatis nudis crenatis radicalibus, caule fubnudo racemofo. · *Syſt. Veg.*
219. *Willd. Sp. Pl.* 1. *p.* 1004. *Reich.* 1.
495. *Jacq. Auſtr. t.* 125. *Pall. It.* 1. *p.* 183.
Scop. Carn. n. 250. *Hort. Kew.* 1. *p.* 237.
Allion. Ped. n. 384. *Gærtn. Fruĉt.* 1. *p.* 262.
t. 55.
VERBASCUM flore cæruleo vel purpureo. *Bauh. Hiſt.* 3.
p. 875.
BLATTARIA perennis, flore violaceo. *Morif. Hiſt.* 2.
p. 497.
BLATTARIA purpurea. *Bauh. Pin.* 241. *Raii Hiſt.* 1096.
BLATTARIA flore purpureo. *Park. Hiſt.* 64. *Ger. emac.* 776, 2.

The Mulleins are all fhewy plants; this fpecies, a native
of the fouthern parts of Europe, having bright purple flowers
is very ornamental, and has been long thought worthy of
cultivation, being feen in our gardens before the time of
GERARD. Is a perfeĉtly hardy perennial, " the roote (as
PARKINSON obferves) abiding fundry yeares," though fome
have fuppofed it to be only biennial, an error ſtill handed
down in MARTYN's MILLER's Diĉtionary. May be eafily
propagated by parting its roots or by feeds, which however
with us it rarely produces, though in fome years abundantly.
Succeeds beſt in a fandy loam with an eaſtern expofure; its
ſtems, if not tied up, are liable to fuffer from high winds.
Blooms through the months of May and June.

Syd. Edwards del.   Pub by T Curtis, S.ᵗ Grescent Nov 1.1805   F.Sansom sculp

Cal. communis polyphyllus ; proprius duplex, fuperus.
Recept. paleaceum f. nudum.

SCABIOSA *caucafea ;* corollulis quinquefidis radiantibus,
    foliis lanceolatis utrinque attenuatis hifpidis bafi
    connatis, calycibus internis externos bis fuper-
    antibus.

DESCR. *Stem* erect, fimple, round, pubefcent, terminating in a
long, naked, round peduncle. *Leaves* oppofite, lanceolate,
narrowed at both ends, quite entire or with here and there a
fmall tooth, hifpid with white, adpreffed, ftiffifh hairs. *Flower*
folitary, very large, radiated. *Involucre* about ten-leaved :
leaflets lanceolate-acuminate, terminated in a fharp mucro,
hairy. *Paleæ* linear-lanceolate, very hairy, longer than the
florets of the difk. *Florets* of the radius tubular, with an unequal five-cleft limb, the three outer fegments many times
larger than the two inner, all obtufe and villous without :
florets of the difk tubular, with a five-cleft equal border, external calyx membranous, cupped, plicate ; internal calyx of
five, briftle-fhaped, black leaflets, twice the length of the external. *Germen* covered with long white down. *Style* oblique.
Stigma globofe. *Seeds* hairy, crowned with both the calyces.
    This fpecies has very great affinity to SCABIOSA *grami-
nifolia,* but the ftem is more erect, the flowers are much
larger, the leaves broader and lefs filvery, with longer and
                        more

more rigid hairs ; but the moſt material difference appears to be in the length of the internal calyx, which in *graminifolia* is hardly longer than the external.

The flower exceeds in ſize that of any other known ſpecies of Scabious, and continues long in beauty.

Raiſed by Mr. Loddiges from ſeeds received by him from Mount Caucaſus. Is a hardy perennial. Flowers in July and Auguſt.

Syd. Edwd del   Pub. by T Curtis, S Geo Crescent No 1 1805   F Sansom sculp

Cal. o. *Cor.* 4—6-petala. *Nectaria* (f. petala interna) 4, fpathulata, petalis alternantia. *Filam.* plurima : exteriora di- latata fubantherifera. *Sem.* ariftata : ariftis pilofis.

Specific Character and Synonyms.

ATRAGENE *americana ;* foliis quaternis ternatis : foliolis cordatis integerrimis, nectariis acutis.

At No. 530 of this work we have figured and defcribed the ATRAGENE *auftriaca*, of which the *alpina* from Siberia has been generally confidered as a variety ; but we there obferved that thefe plants were probably diftinct fpecies. We are now able to afcertain that they really are fo, and to add a third, a native of North-America, which, with *ochotenfis* of PALLAS, makes up the whole of the fpecies from which the above ge- neric character is formed ; ATRAGENE *capenfis* and all the other fpecies mentioned by WILLDENOW, except perhaps ATRAGENE *zeylanica*, probably do not belong to this genus ; which is chiefly diftinguifhed from CLEMATIS by the prefence of the nectaries or internal petals, and by its very fingular manner of growth ; every gemma (to which there appears to be nothing fimilar in CLEMATIS) producing as it were a diftinct plant, confifting of two or four leaves, with a peduncle bearing a folitary flower in the centre. Thefe plants are connected together by farmentous ftalks, but on very elevated mountains the ATRA- GENE *auftriaca* is entirely deftitute of thefe ftalks, and the whole plant confifts merely of two radical leaves with a folitary flower,

flower, fupported on a fcape. It was in this form only that the plant had occurred to HALLER, at the time he wrote his *Hiftoria Stirpium Helvetiæ*; probably alfo LINNÆUS had not feen it in any other, when he defcribed the leaves as radical, and called the peduncle a fcape : and even in cultivation the feedling plants will fometimes flower before any running fhoot appears.

The nectaries or internal petals have been hitherto faid to be numerous, but we conftantly find four that are fomewhat different from the reft, placed alternately with the petals and without any veftige of anthers ; all the others, generally about twelve, having more or lefs appearance of anthers at their tips, we confider as dilated filaments. The nectaries afford an excellent mark of difcrimination between ATRAGENE *auftriaca* and *fibirica,* which are otherwife not eafily diftinguifhed by words, thofe of the latter being emarginate or linear-obcordate, whereas thofe of the former are quite entire at the point.

Having had an opportunity of feeing three fpecies flower at Mr. LODDIGES, at Hackney, this fummer, we have not omitted comparing them together, and think they may be fafely concluded to be diftinct and characterized as follows :

1. ATRAGENE *auftriaca* ; foliis binis duplicato-ternatis: foliolis ovatis ferratis, nectariis obtufis.

2. ATRAGENE *fibirica* ; foliis binis duplicato-ternatis: foliolis ovatis ferratis, nectariis emarginatis.

3. ATRAGENE *americana* ; foliis quaternis ternatis: foliolis cordatis integerrimis, nectariis acutis.

Of the fourth fpecies, the *ochotenfis* of PALLAS, we know nothing but the little this author has given of it in his Flora Roffica, vol. 2, p. 69. It has fix petals, but in other refpects has the habit of the reft.

According to JUSSIEU, what we have denominated corolla is a calyx, and our nectaries are petals. We prefer keeping to the Linnæan terms, and have only adopted that of nectaries inftead of his internal petals, in conformity to his own language in the reft of the order.

The plant now figured flowers nearly at the fame time with ATRAGENE *auftriaca,* a month later than *fibirica,* is hardly lefs ornamental, and has the exclufive advantage of being agreeably fcented. Was raifed from feeds from North-America, by Mr. LODDIGES ; appears to be perfectly hardy, and to produce feeds freely, by which it may be propagated without difficulty, and makes a very defirable addition to our climbing fhrubs.

L.Edwards del Pub by T.Curtis St Geo Crefcent Nov.r 1.1805. F.Sanfom sc.

EPIDENDRUM SINENSE. CHINESE EPIDENDRUM.

✱✱✱✱✱✱✱✱✱✱✱✱✱✱✱✱✱✱

Clafs and Order.

GYNANDRIA DIANDRIA.

Generic Charaĉter.

Neĉarium undulatum, obliquum, reflexum.

Specific Charaĉter and Synonyms.

EPIDENDRUM *finenfe ;* foliis enfiformibus nervofo-ftriatis
radicalibus, floribus nutantibus, petalis fub_
æqualibus, neĉario revoluto maculato fubtus
concavo, braĉtea germine parum breviore.
EPIDENDRUM *finenfe. Bot. Repof.* 216. *Donn. Cantab.* 166.

This plant belongs to the genus CYMBIDIUM of SWARTZ[*],
and is very nearly allied to EPIDENDRUM *enfifolium* of LIN-
NÆUS, figured by Dr. SMITH, in his *Spicilegium Botanicum ;*
fo nearly indeed that, perhaps, fome may be inclined to confider both as varieties. It is a larger plant, the leaves wider and
more evidently nerved, the flowers larger, darker coloured,
and more nodding ; the braĉte below each flower is above
two-thirds the length of the germen, whereas in *enfifolia* it is
fcarcely one-third the length ; and the germen is much curved,
which in *enfifolia* is nearly ftraight. But even thefe diftinĉions,
flight as they are, we can hardly infift upon, unlefs we had
feen more fpecimens of both in flower : there is however a
confiderable difference in the general appearance, and cultivators think them diftinĉt. We have a drawing of the other

---

[*] See the Profeffor's paper on the genera of ORCHIDEÆ, in *Traĉts relative to
Botany*, which we are informed was tranflated from the Swedifh language, by
our friend Mr. CHARLES KÖNIG, two years prior to the publication.

plant

plant by us, and perhaps, when we publish this, we may be able to fpeak more decidedly upon the fubject.

A native of China, from whence it was introduced by the late Mr. SLATER ; has been hitherto treated as a ftove plant, but does not require fo much heat as the Weft-Indian fpecies, thriving luxuriantly in the confervatory.

The genus EPIDENDRUM, as at prefent conftituted, certainly contains many very heterogeneous fpecies; but, perhaps, until a much larger number of them have been figured and defcribed, it may be better to fuffer them to remain as they are. Profeffor SWARTZ has done much, but we acknowledge that we are deterred from following his arrangement, by obferving feveral fpecies united which can hardly belong to the fame genus ; thus EPIDENDRUM *cucullatum*, figured above, No. 543, ranks with our prefent plant under CYMBIDIUM!

Syd. Edwards del. Pub by T Curtis, St Geo Crescent Dec 1 1805

ONOSMA TAURICA. GOLDEN-FLOWERED ONOSMA.

Clafs and Order.

PENTANDRIA MONOGYNIA.

Generic Chara&er.

Cor. campanulata: fauce pervia. *Semina* 4.

Specific Chara&er and Synonyms.

ONOSMA *taurica ;* caulibus fimplicibus e bafi multicipe, foliis lineari-lanceolatis utrinque albo-pilofis, fru&ibus ere&is. *Marfch. v. Biberflein Terek u. Kur. p.* 138.

ONOSMA *taurica. Pallas Tableau de la Tauride, p.* 47. *Annals of Bot. v.* 2. *p.*

This plant is not entirely new in our gardens, where it has generally paffed for ONOSMA *echioides* of LINNÆUS, a much larger plant, greatly branched, clothed with very long yellowifh hairs, and having entirely the habit of ECHIUM *vulgare.* We at firft fufpe&ed it to be the ONOSMA *fimpliciffima ;* but, from the confufion in the fynonymy and the want of precifion in the fpecific chara&ers, it was not eafy to determine the queftion. But fortunately in our fearch we met with fpecimens exa&ly corresponding with our plant, in a colle&ion fent from Caucafus to Sir JOSEPH BANKS, by Count MUSCHIN PUSCHKIN, under the name which we have adopted ; and with the affiftance of our kind friend Mr. CHARLES KONIG, we are enabled to give the fpecific chara&er as drawn up by Marfchal v. BIBERSTEIN, by whom we are informed that it is frequent in the open hills of Tauria, about Karaffubafac and Sympheropolis, and alfo in the mountains of the Cafpian Caucafus, flowering in May and June. A careful examination of the dried fpecimens left us almoft without doubt, yet the obfervation of this author, that

the

the flowers are of a full yellow colour, affords an additional proof of the identity of the plants.

It is a hardy perennial, but requires the fame care as moft other alpine plants, which are often preferved with more difficulty through our moift winters and variable fprings, than the natives of warmer climes.

Our drawing was taken at the nurfery of Meffrs. WHITLEY and BRAME, Old-Brompton, in June laft.

J. Edwards del   Pub by. T.Curtis, St Geo Crescent Dec 1.1805   F. Sansom sculp

GOODENIA GRANDIFLORA. LARGE-FLOWERED GOODENIA.

****

Class and Order.

PENTANDRIA MONOGYNIA.

Generic Character.

Capf. 2-locularis, 2-valvis, polyfperma, diffepimento parallelo, *Sem.* imbricata. *Cor.* fupra longitudinaliter fiffa, genitalia ex-ferens: limbo 5-fido, fecundo. *Antheræ* lineares, imberbes. *Stigma* urceolatum, ciliatum. SMITH.

Specific Character.

GOODENIA *grandiflora;* caule herbaceo angulato, foliis cordatis dentato-ferratis villofis: inferioribus pinnatis, floribus axillaribus ternis, capfulis pentagonis gibbis.

DESC. *Root* annual or biennial, fibrous. *Stalk* branched, three or four feet high, fix or feven angled, and deeply fur-rowed, hairy, filled with light pith like elder. *Leaves* alternate, on long petioles; upper ones fimple; heart-fhaped, acuminate, fawed, with teeth nearly perpendicular, foft and fomewhat clammy; lower ones pinnated, the terminal leaflet the fame as the upper leaves and much larger than the others. *Flowers* grow generally by threes, but at the upper part of the plant frequently folitary, from the axils of the petioles, the common peduncle very fhort or almoft none, with a fhort fubulate braĉte at the bafe of each pedicle. *Calyx* fuperior or growing to the germen, divided into five fubulate fegments, perfiftent. *Corolla* yellow, irregular; laciniæ 5, ovate-lanceolate, three-nerved on the under furface, margin undulated; the claws of the three lower ones adhere together, but the two upper
laciniæ

GOODENIA GRANDIFLORA. LARGE-FLOWERED GOODENIA.

✶✶✶✶✶✶✶✶✶✶✶✶✶✶✶✶✶✶✶

Class and Order.

PENTANDRIA MONOGYNIA.

Generic Character.

Capf. 2-locularis, 2-valvis, polyfperma, diffepimento parallelo. *Sem.* imbricata. *Cor.* fupra longitudinaliter fiffa, genitalia exferens: limbo 5-fido, fecundo. *Antheræ* lineares, imberbes. *Stigma* urceolatum, ciliatum. SMITH.

Specific Character.

GOODENIA *grandiflora;* caule herbaceo angulato, foliis cordatis dentato-ferratis villofis: inferioribus pinnatis, floribus axillaribus ternis, capfulis pentagonis gibbis.

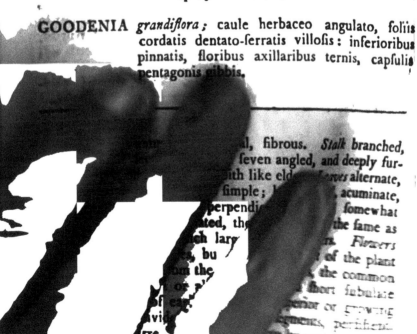

, fibrous. *Stalk* branched, feven angled, and deeply furth like eld leaves alternate, fimple; acuminate, perpendi fomewhat ted, th the fame as ch lar. Flowers bu of the plant om the the common or a fhort fubulate ea ior or growing vi ments, perithent rre laciniæ are diftinct and erect, forming a hood or umbrella over the ftyle. *Stamens* 5, quickly perifhing; filaments fhort, recurved, inferted into the crown of the germen; anthers linear. *Style* erect, incurved, hairy; *Stigma* large, compreffed, ciliated at the mouth. *Capfules* five-angled, gibbous, two celled, fplitting at the point into four valves, to difcharge the feeds. *Seeds* lenticular, attached to the diffepiment, which is detached from the fides of the capfule at its upper part. Embryo in the centre of a flefhy perifperm : radicle defcendent.

As foon as the flower opens, the ftamens are bent quite away from the ftigma ; but the anthers in this genus, as in moft, if not all, the family of campanulaceæ, fhed their pollen before the corolla is expanded. If the flower-bud in this fpecies be carefully opened a day or two before its proper feafon of expanding, a moft curious fpectacle offers itfelf, the ftigma will be found erect, open, in the fhape of a cup, and fometimes completely filled with the pollen, fhed from the anthers, which now connive over its mouth. Before the flower opens, the ftyle is much lengthened, and the ftigma clofes, the filaments at the fame time fhrinking away.

The flowers have a fweet, but not very agreeable, fmell, and the whole plant partakes of the fame.

If brought forward by fowing the feeds in a gentle hot-bed in the fpring, and the young plants be afterwards fet out in open ground, they will flower about the beginning of Auguft, and continue to bloffom till the froft fets in. The feeds will ripen in October. Requires a plentiful fupply of water.

We received this plant from Mr. WHITLEY, of Old-Brompton, who raifed it from feeds from New South-Wales.

Syd. Edwards del. Pub. by T Curtis, St Geo Crescent Dec 1 1805. F. Sansom sculp.

PIMELEA LINIFOLIA. FLAX-LEAVED PIMELEA.

****

Class and Order.

DIANDRIA MONOGYNIA.

Generic Character.

Cal. o. *Cor.* 4-fida. *Stam.* fauci inferta. *Nux* corticata, 1-locularis. SMITH.

Specific Character.

PIMELEA* *linifolia;* foliis lineari lanceolatis, capitulis terminalibus involucratis, corolla extus villofa. *Smith Nov. Holl.* 1. *p.* 31. *t.* 11. *Willd. Sp. Pl.* 1. *p.* 50. *Mart. Mill. Dict.*

DESC. *Stem* fhrubby with a ferrugineous bark : branches erect, fubdichotomous. *Leaves* oppofite, decuffate, horizontal, fmooth, linear-lanceolate, thickened at the margin : middle nerve depreffed underneath, raifed on the upper furface. *Inflorescence* a terminal umbel, fupported by an involucre of four ovate, acute, quite entire leaflets. *Flowers* white, fcentlefs. *Corolla* tubular, with a quadrifid limb; tube and the two exterior laciniæ of the limb villous, the two interior fmooth. *Filaments* two, inferted into the margin of the tube, the length of the laciniæ. *Anthers* orange-coloured. *Germen* fuperior, oval, green : *ftyle* filiform, longer than the tube (in the centre flowers fhorter): *ftigma* fimple.

For this elegant fmall greenhoufe fhrub, which continues to bloom through the greateft part of the year, we are indebted to Mr. LODDIGES of Hackney.

It is a native of New South-Wales; may be propagated by cuttings or from feeds imported from its native country.

* For what reafon Dr. SOLANDER applied this name to the genus is unknown; but, being derived from πιμελή, *fat,* in pronunciation the accent fhould be laid upon the laft e, not on the firft ;—Pimeléa not Piméléa, as has been duly noticed by Prof. MARTYN.

raifed it from feeds fent by Lady Gwillim from Madrafs, under the name of the Seringapatam Hollyhock.

We neglected to mention in our laft number that the opportunity of giving a figure of the Hibiscus *paluftris* (No. 882) was afforded us by the fame ingenious cultivator, who, obferving that under the ordinary treatment this plant never flowered, removed it from the open ground into the ftove towards the end of the fummer 1800, and thus forced it into bloffom ; a practice that has been fince fuccefsfully followed by others.

HEXANDRIA MONOGYNIA.

*Generic Charaɛler.—Vid. N*ᵘᵐ· 800.

Specific Charaɛler and Synonyms.

LILIUM *Martagon* foliis ſtriɛtiuſculis, obovato-lanceolatis, nudis, ſubcorrugato-venoſis, deorſum remote (ſæpe etiam duplicatim) verticillatis ; racemo multifloro ; corollis pendulo-cernuis, laciniis revoluto-reflexis, intimis apice villoſis atque obtuſiſſimis. G.

LILIUM *Martagon. Hort. Cliff.* 120. *Gærtn. Sem.* 2. *p.* 17. *t.* 83. *Flor. Arragon.* 301. *Jacq. Auſtr. t.* 351. *Willd. Sp. Pl.* 2. 88. *Hort. Kew.* 1. *p.* 431.

LILIUM foliis verticillatis floribus pendulis revolutis. *Hall. Helv. n.* 1233.

LILIUM ſilveſtre ſive montanum. *Cluſ. Hiſt.* 133. *Dod. Pempt.* 201.

LILIUM floribus reflexis montanum. *Bauh. Pin.* 77.

MARTAGON. *Camer. Epit.* 617.

MARTAGON Imperiale. *Park. Parad.* 28.

MARTAGON Imperiale moſchatum. *Hort. Eyſt. Plant. Æſtiv. Ord. II. necnon tab.* 2 *ſequent.*

A native of the Auſtrian mountains and of ſome other parts of Germany. Becomes in our gardens a large plant, between three and four feet high, or more. Corolla gloſſy, glazed like porcelain, varies in its colour from purple, to whitiſh purple and white. Leaves coarſe and harſh. Braɛtes often double. Diſtance between the whorls of leaves about the length of the leaves. Pedicles long, aſcendently patent. Stamens far ſhorter than the corolla. Style clubbed, twice longer than germen. Perfeɛtly hardy ; and flowers about July or Auguſt.

There are two Auſtrian ſpecimens in the Bankſian Herbarium, the one with a naked, the other with a pubeſcent ſtem ; the former is figured in JACQUIN, and is a ſlenderer ſmaller plant, but we can hardly think them ſpecifically different. G.

I

Syd.Edwards del. Pub. by T.Curtis. St. Geo Crescent Dec 1 1805. F. Sanf m sculp

HEMEROCALLIS CÆRULEA. CHINESE DAY-LILY.

****

Claſs and Order.

HEXANDRIA MONOGYNIA.

Generic Charaƈter.

Cor. baſi infundibuliformis, limbo campanulato, ſexfido, apice revoluto. *Stam.* declinata. *Stigm.* 3-gonum. *Germen* inferum. *Capſ.* 3-gona, corolla teƈta.

OBS. *Radices faſciculatæ, et flores in ſcapo corymboſi aut racemoſi.* JUSS. *Vix* AGAPANTHO *niſi habitu diſtinguenda.* G.

Specific Charaƈter and Synonyms.

HEMEROCALLIS *cærulea* foliis petiolatis, acuminato-ovatis, ſubplicato-nervoſis; braƈteis pedicellos parum ſuperantibus, ſcarioſo-membranaceis; racemo multifloro; corollæ tubo ſulcato-cylindrico limbum ſubſemi-ſex-partitum campanulatum vix æquante; ſtaminibus apice involutis. *G.*

HEMEROCALLIS *cærulea. Bot. Rep. tab.* 6. *Liliac. a Redouté,* 106. *t.* 106. *Venten. Malmaiſ.* 18. *t.* 18.

HEMEROCALLIS *japonica* β. *Willd. Sp. Pl.* 2. 198.

After the very detailed deſcriptions in the works of RE-DOUTE and VENTENAT (above cited) of this now not uncommon plant, it would be ſuperfluous to add any further deſcription to that contained in the ſpecific charaƈter, which ſeems to diſtinguiſh it from HEMEROCALLIS *japonica,* of which it has been deemed a variety by WILLDENOW.

A native of China; thought to bloom beſt in the ſtove; but thrives very well in a greenhouſe; and ſome cultivators aſſure us, that it ſucceeds in the open ground better than with any other treatment.

Introduced by GEORGE HIBBERT, Eſq. Seeds freely, and is eaſily propagated by offſets.

Our drawing was taken from a ſmall few-flowered ſpecimen. *G.*

Syd Edwards del.   Pub by TCurtis, St Geo Crescent Dec 1 1805   F Sansom sculp

WITSENIA CORYMBOSA. CORYMBOSE WITSENIA.

****

Clafs and Order.

TRIANDRIA MONOGYNIA.

Generic Character.

Inflor. aut fafciculata, involucro bracteis pluribus minoribus quafi imbricatim calyculato, aut paniculata fingulo flore fpatha bivalvi uni-bracteata excepto. *Cor.* regularis, æqualis; vel tubulofa limbo 6-partito, vel tota hexapetalo-partita. *Stam.* erecta, laciniarum bafi adnata. *Stigm.* tria. *Capf.* putaminea, trifariam dehifcens. *Sem.* plura, angulato-preffa. *G. Vid. Ann. of Bot. v.* 1. *p.* 236.

Specific Character and Synonyms.

WITSENIA *corymbofa* inflorefcentia corymbofo-paniculata ; fpatha bivalvi unibracteata ; corolla hypocrateri-formi ; tubo erecto fpatham pluries, limbum ex-planatum parum, excedente; filamentis fubnullis ; piftillo corollam fubfuperante. *G.*

A new fpecies, probably the firft of the genus ever culti-vated in an European garden. Raifed from feeds received from the Cape of Good Hope by Mr. HIBBERT at Clapham. Agrees in herb with the three fpecies enumerated in the *Annals of Botany, v.* 1. *p.* 237, but differs from them widely in its inflorefcence, which is a many-flowered corymbofe panicle, borne on a flat two-edged rachis and pedicles. This is the only genus of the order that has any thing of a frutefcent habit ; the rootftock becomes woody and lengthens into a flat ftem, covered with enfiform flabellately diftich leaves, which are im-bricately equitant towards their bafes ; this ftem is analogous to the rootftock of the IRIS, but is lignefcent, flender, and upright, inftead of being flefhy, thick, and procumbent. The prefent plant appears to be the connecting link of its genus with LAPEYROUSIA, fee plate 595, and probably on its other fide alfo with ARISTEA, fee A. *cyanea*, No. 458. Requires the protection of a greenhoufe; flowers in September; fcentlefs. *G.*

Syd Edward del   Pub by T Curtis St Geo Crescent Jan 1800 F Sansom sculp

DIADELPHIA DECANDRIA.

Generic Charaƈter.

Vexilli baſis callis 2, parallelis, oblongis, alas ſubtus comprimentibus.

Specific Charaƈter and Synonyms.

DOLICHOS *Lablab ;* volubilis, leguminibus ovato-acinaciformibus, ſeminibus ovatis hilo arcuato verſus alteram extremitatem. *Sp. Pl.* 1019. *Reichard* 941. *Willd.* 3. *p.* 1037. *Roy. Lugdb.* 368. *Hort. Upſ.* 214. *Haſſelquiſt.* 483. *Eng. Edit.* 252. *Mart. Mill. Diƈt.* 2. *Hort. Kew.* 3. *p.* 31. *Gært. Fruƈt.* 2. *p.* 322. *t.* 150.

PHASEOLUS peregrinus 14 Leblab. *Cluſ. Hiſt.* 2. *p.* 227.

PHASEOLUS ægpytius nigro ſemine. *Bauh. Pin.* 341. *Raii Hiſt.* 888.

PHASEOLUS. *Riv. Tetr. t.* 29. *f.* 4.

PHASEOLUS niger Lablab. *Alp. Ægypt.* 74. *t.* 75. *? Veſt. Ægypt.* 27. *?*

Although always conſidered as a native of Egypt, HASSELQUIST aſſures us, that this plant is only cultivated there, and was moſt probably introduced from Europe, as it is called by the inhabitants the European Bean. Is cultivated for the table in ſeveral warm countries, in the ſame manner as the kidney-bean is with us ; indeed Phaſeolus and Dolichos are very nearly allied. Is uſually conſidered as a ſtove plant, but is marked by Mr. DONN as a hardy annual.

ALPINUS deſcribes his Lablab as a climbing evergreen tree, as large as a vine, enduring a hundred years or more, and in frequent uſe in the gardens of Egypt for making ſhady bowers. Surely this muſt render it very doubtful if his plant be the ſame as ours; more eſpecially as he deſcribes the pods as being long.

Flowers from July to September. Said in the Hortus Kewenſis to have been cultivated by the Ducheſs of BEAUFORT, in 1714. Communicated by Mr. GIBBS, Seedſman, Piccadilly.

Syd Edwards del   Pub by T Curtis. S.t G.eo Crefcent Jan 1 1800   Sanfom sculp

PYROLA MACULATA. SPOTTED-LEAVED WINTER-GREEN.

Clafs and Order.

DECANDRIA MONOGYNIA.

Generic Character.

Cal. 5-partitus. *Petala* 5. *Capf.* 5-locularis, angulis dehifcens.

Specific Character and Synonyms.

PYROLA *maculata ;* pedunculis fubbifloris, foliis lanceolatis dentato-ferratis variegatis, ftigmate fubfeffili hemifphærico.

PYROLA *maculata ;* pedunculis bifloris. *Spec. Plant.* 567. *Reich.* 2. *p.* 300. *Willd.* 2. *p.* 622. *Hort. Kew.* 2. *p.* 34, *Mart. Mill. Dict. n.* 5.

PYROLA *maculata ;* foliis lanceolatis, rigide ferratis, fafcia longitudinali difcolore notatis : fcapo bi- five rarius trifloro ; filamentis lanuginofis : ftigmate feffili. *Michaux Flor. Bor. Amer.* 1. *p.* 251.

PYROLA petiolis apice bifloris vel trifloris. *Gron. Virg.* 48.

PYROLA marilandica minor folio mucronato arbuti. *Pet. Muf.* 675.

PYROLA Mariana, arbuti foliis anguftioribus, trifoliata ; ad medium nervum linea alba utrinque per longitudinem difcurrente. *Pluk. Mant. p.* 157. *t.* 348.

This fpecies has very near affinity with the PYROLA *umbellata*, No. 778, and is likewife a native of North-America. The leaves are more rigid, pointed, and marked with a white line fometimes only on each fide the midrib, but more ufually

branching

branching out with the veins alfo; the ftem is twifted and the leaves, though oppofite, are generally turned to one fide and crowded towards the upper part; the petals are more reflexed and the bafe of the filaments orbicular and deeply fringed. The ftem is faid in MILLER's Dictionary to be a foot and half high, but we have never feen it more luxuriant than in the fpecimen from whence our drawing was taken, with which we were favoured by Mr. LODDIGES of Hackney. Requires the fame treatment as PYROLA *umbellata.*

Cultivated by PHILIP MILLER in 1759. Flowers in June and July, and continues long in bloffom.

Syd.d Edwards del   Pub. by T Curtis St Geo. Crescent Jan. 1 1806   F. Sansom sculp

Orontium Japonicum. Japan Orontium.

****

Class and Order.

Hexandria Monogynia.

Generic Character.

Spadix cylindricus, tectus flosculis. *Cor.* 6-partitæ nudæ. *Stylus* nullus. *Folliculi* 1-spermi.

Specific Character and Synonyms.

ORONTIUM *Japonicum* rhizomate oblongo, crasso, carnoso; fibris crassis; foliis lanceolatis, nervoso-striatis, basi convolutis; scapo aroideo, istis aliquoties breviore; spica ovali-oblonga favosim conferta; baccis obovato-olivæformibus rubentibus. *G.*

ORONTIUM *Japonicum. Thunb. Jap.* 144. *Syst. Veg.* 350. *Hort. Kew.* 1. 474. *Sp. Pl.* 2. 200.

KIRO et RIRJO, vulgo OMOTTO. *Kæmpf. amæn. exot.* 785. *Banks. Ic. Kæmpf. tab.* 12. *optima.*

This singular plant is a native of Japan, and has been admirably described as well as figured by the indefatigable Kæmpfer, who says the Japanese plant it to cover by its luxuriant leaves the waste spots and corners of their pleasure gardens. The root is rather bitter, but without the acritude of that of Arum; and consists of an elongated, subcylindric, thick, fleshy root-stock with fleshy fibres; leaves radical, opposite, lanceolate, several, convolute at their bases; from one to two feet long, one to three inches broad, deep green scape several times shorter than these, one to three inches high, spike oval-oblong, thick set with sessile flowers, as crowded as the cells of a honey-comb, parted by membranous bractes, corollas campanulate with the points of the segments somewhat inflected; the berries are said by Kæmpfer to be about the size and form of a small olive, containing a scarlet pulp, with a single heart-shaped seed of a bony substance; have a nauseous taste. Thunberg's description of this plant is inconceivably defective and erroneous; he

says

fays he found it near Nagafaki as well as in feveral other parts of the country; if he had not quoted KÆMPFER and fent a fpecimen, which is preferved in the Bankfian Herbarium, it would have required no fmall fhare of fagacity to have difcovered what plant he meant,

Introduced into this country by Mr. GRÆFER in 1783. Ufually cultivated in the dry ftove; when it flowers about January; but is fufficiently hardy to endure our ordinary winters in the open air, and flowers from March to June; we have feen it in feveral collections; our drawing was made at that of Mr. MALCOLM's, Kenfington; propagated by off-fets. We loft the opportunity of examining the flowers, fo that we have trufted to a dried fpecimen and the drawing for their defcription.

The above effential character we have added as we found it in the books; but think that it is no ways adapted to the prefent fpecies; nor indeed to ORONTIUM *aquaticum*, as far as we can make out from the dried plant. *G.*

Syd Edwards del. Pub. by T. Curtis St Geo. Crescent Jan 1 1806 F. Sansom sc.

CONVALLARIA RACEMOSA. CLUSTER-FLOWERED SOLOMON'S SEAL.

****

Clafs and Order.

HEXANDRIA MONOGYNIA.

Generic Character.

Cor. vel fexfido-tubulofa, vel globofa, aut fexpartito-patens. *Stigma* trigonum. *Bacca* fupera, 3-locularis, ante maturitatem maculata.

OBS. *Species una foliis 6-verticillatis, altera radicalibus vaginantibus fpathaceis, cætera feffilibus alternis.* JUSSIEU.

Specific Character and Synonyms.

CONVALLARIA *racemofa ; herba tota pubefcens ;* foliis ovali-lanceolatis, longe acuminatis, feffilibus, alternis caulinis ; racemo compofito, terminali ; corollis rotatis, parvis, crebris ; laciniis peranguftis. G.

CONVALLARIA *racemofa. Sp. Pl.* 452. *Hort. Cliff.* 125. *Gron. Virg.* 38, 52. *Horl. Kew.* 1. 455. *Mart. Mill. Dict. Willd. Sp. Pl.* 2. 162. *Michaux Fl. Bor-Amer.* 1. 202.

POLYGONATUM racemofum. *Corn. Canad.* 36. *t.* 37. *Morif. Hift. f.* 13. *t.* 4. *f.* 9. *Park. Theat.* 697. *f.* 8.

___

A native of Virginia and Canada ; MICHAUX fays he alfo found it on the mountains in Carolina ; there is a diminutive ftrongly pubefcent variety in the Bankfian Herbarium from the neighbourhood of New-York ; from which fource we alfo learn that it is called in the language of the Cherokee Indians,. *Oiole Nowote* (Child's Phyfic).

Cultivated

Cultivated in this country by Mr. John Tradescant, jun. in 1656. A hardy perennial, growing to the height of about two feet; leaves ribbed, lower ones not unlike thofe of Plantain. Blooms about June; corollas minute, fucceeded by fmall red berries. Eafily propagated by parting its roots; fucceeds beft in a light foil and fhady fituation.

Ranks among the *Smilaces*, a denomination given to a fub-divifion of this genus, including the fpecies which have rotate corollas. *G.*

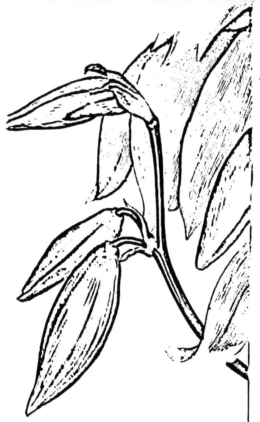

Pub by T. Curtis S.t Geo Crescent Jan 1 1.....

****

Clafs and Order.

HEXANDRIA MONOGYNIA.

Generic Charaƙer.

Cor. hexapetalo-partita, campanulata. *Stam.* filamentis fu-
perne craffioribus ; antheris minimis. *Stigm.* 3, feffilia. *Capf.*
(Bacca carnofa *Gærtn.)* obtufe trigona. *Sem.* plana.

Specific Charaƙer and Synonyms.

YUCCA *filamentofa (acaulis)* foliis oblongo-lanceolatis, mar-
     ginibus diftanter filiferis ; ftigmatibus recurvato-
     patentibus. G.
YUCCA *filamentofa. Syft. Vegetab. Murr.* 337. *Mill. Diƌ.* 4.
     Gron. Virg. 152 ; 58. *Trew Ebret. t.* 37. *Hort.*
     Kew. 1. 465. *Willd. Sp. Pl.* 2. 184. *Michaux Flor.*
     Bor-Amer. 1. 196.
YUCCA foliis filamentofis. *Morif. Hift.* 2. *p.* 419.
YUCCA virginiana foliis per marginem apprime filatis. *Pluk.*
     Alm. 396.

According to MICHAUX a native of the weftern parts of
Carolina and Virginia, growing on wilds near the fea-fhore,
with a ftem rifing fometimes to the height of five feet.
An old inhabitant of our gardens, having been cultivated
in them as far back as 1675. Hardy. Very ornamental ;
flowers about September or Oƌober. Is not uncommon in
our Nurferies ; propagated by fuckers. G.

E R R A T A.

No. 854, l. 14, pro " LACHENALIA anguftifolia," lege " LACHENALIA
lanceæfolia."

No. 895, l. 26, for " a flat two-edged rachis and pedicles," read " a flat
two-edged ftalk and rachis."

Syd Edwards del   Pub by T Curtis St Geo Crescent Jan 1 1806   F Sansom sculp

LOBELIA ERINUS. ASCENDING LOBELIA.

❋❋❋❋❋❋❋❋❋❋❋❋❋❋❋❋❋❋

Claſs and Order.

PENTANDRIA MONOGYNIA, *olim* SYNGENESIA MONOGAMIA.

Generic Charaƈer.

Cal. 5-fidus. *Cor.* 1-petala, irregularis. *Antheræ* cohærentes. *Capſ.* infera 2- ſeu 3-locularis.

Specific Charaƈer and Synonyms.

LOBELIA *Erinus ;* caulibus filiformibus tortuoſo-ereƈis, foliis obovatis inciſo-dentatis glabris petiolatis, floribus racemoſis terminalibus, capſulis bilocu-laribus.

LOBELIA *Erinus ;* foliis lanceolatis ſerratis glabris, caule flexuoſo ereƈo, pedunculis axillaribus folio lon-gioribus. *Thunb. Prod.* 40. *?*

LOBELIA *Erinus. Syſt. Vegetab.* 802. *? Willd. Sp. Pl.* 1. *p.* 948. *?*

There is ſo much difficulty in determining many ſpecies of Lobelia, that it is not without heſitation that we give this plant, as the LOBELIA *Erinus ;* it ſeems probable indeed that LINNÆUS himſelf took up different plants, under this name. At firſt ſight there appears to be a great ſimilarity between the preſent plant and that figured at No. 514 of this work, but a nearer examination ſhews a ſufficient difference ; the flowers hardly differ, except that in the *Erinus* the flower ſegments are more obtuſe and the colour is more brilliant ; the ſtalks, though weaker, affeƈ a more upright growth and ſeem cal-culated to riſe up among graſs ; the whole plant is ſmooth, whereas the other is covered in every part with ſtiffiſh hairs ; the root is perennial, as indeed it is in *bicolor*, though ſaid by us, erroneouſly, to be annual ; the capſules of both are two-celled, in which reſpeƈ our plant does not correſpond with the deſcription of LINNÆUS.

We

We were favoured with the plant from which our drawing was made by the lady of J. WILSON, Efq. of Iflington.

Native of the Cape of Good Hope; flowers from June to September; fhould be kept in an airy part of the greenhoufe during the winter, and requires a plentiful fupply of water whilft in bloom.

The LOBELIA *bicolor*, No. 514, as we at firft fufpeded, is perhaps a mere variety of the LOBELIA *pubefcens* of the Hortus Kewenfis; the flowers of the former frequently degenerate to white, yet we have never feen them of fo pure a white as in the original *pubefcens*, in which too the ftalks are more ered, more branched at the upper part, and the leaves are broader, more toothed, and fomewhat more pubefcent.

Syd Edwards del.   Pub by T Curtis, St Geo Crescent 1.1805   F.Sansom

ANTIRRHINUM ASARINA. HEART-LEAVED SNAP-DRAGON.

****

Class and Order.

DIDYNAMIA ANGIOSPERMIA.

Generic Character.

Cal. 5-phyllus. *Corollæ* bafis deorfum prominens, nectarifera. *Capf.* 2-locularis.

Specific Character and Synonyms.

ANTIRRHINUM *Afarina ;* foliis oppofitis cordatis crenatis, corollis ecaudatis, caulibus procumbenti- bus. *Spec. Pl.* 860. *Reich.* 3. *p.* 139. *Willd.* 3. *p.* 259. *Hort. Cliff.* 313. *Hort. Kew.* 2. *p.* 338. *Mart. Mill. Dict. n.* 43.
ASARINA *procumbens. Mill. Dict.*
ASARINA. *Lob. Ic.* 601. *Bauh. Hift.* 3. *p.* 856. *Hort. Rom.* 3. *t.* 3.
ASARINA galeata. *Morif. Hift.* 3. *p.* 432. §. 11. *t.* 21. *f. penult.*
HEDERA faxatilis magno flore. *Bauh. Pin.* 306. *Ger. Emac.* 856. 2. *Raii. Hift.* 567.

This fpecies of Antirrhinum, although cultivated by BOBART, at Oxford, as long ago as the year 1699, and again by PHILIP MILLER, in the Phyfic Garden at Chelfea, before 1748, may be neverthelefs confidered as a rare plant.

Native of Italy, and, according to MILLER, a hardy an-nual, faid in the Kew Catalogue to be perennial and hardy ; but Mr. DONN, of Cambridge, juftly confiders it as a greenhoufe plant.

Flowers from July to September. Our drawing was taken from a plant communicated by NAPIER and CHANDLER, Nurferymen, Vauxhall.

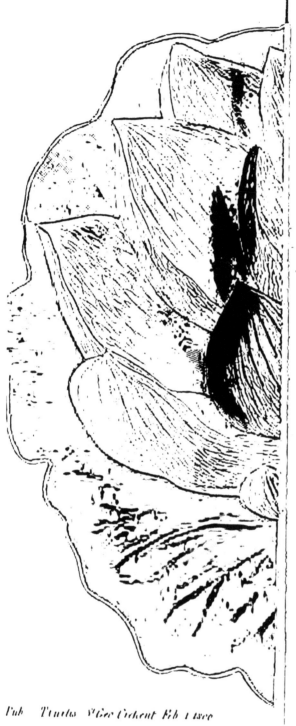

Pub. Trubrs S Geo Cochent Feb 1 1800

Syd. Edwards del   Pub by T Curtis, St Geo Crescent Feb 1 1806   F Sansom sculp

NELUMBIUM SPECIOSUM. SACRED BEAN OF INDIA.

****

Clafs and Order.

POLYANDRIA POLYGYNIA.

Generic Charaƈter.

Cal. 4 feu 5-phyllus. *Cor.* polypetala. *Nuces* 1-fpermæ, ftylo perfiftente coronatæ, receptaculo truncato immerfæ.

Specific Charaƈter and Synonyms.

NELUMBIUM *fpeciofum ;* foliis peltatis, orbiculatis integer-
    rimis, pedunculis petiolifque. muricatis, co-
    rolla polypetala. *Willd. Sp. Pl. v.* 2. *p.* 1258.
NELUMBO *nucifera. Gært. Fruƈ.* 1. *p.* 73. *t.* 17. *f.* 2.
NYMPHÆA *Nelumbo ; Sp. Pl.* 730. *Reich.* 2. *p.* 579.
    Lerche in Nov. Aƈ. Nat. Cur. v. 5. *app.*
    p. 191. *Thunb. Jap.* 223. *Lour. Cochinch.*
    340. *Pluk. Alm.* 267. *Phyt. t.* 322. *f.* 1.
CYAMUS *Nelumbo. Smith Exot. Bot. t.* 31 & 32.
CYAMUS *myfticus. Salifb. in Ann. of Botany, v.* 2. *p.* 73.
TARATTI. *Rumph. Amb.* 6. *p.* 168. *t.* 73.
TAMARA. *Rheede Malab. v.* 11. *p.* 59. *t.* 30.
REN. *Kæmpf. Amæn. Exot.* 880.

In the courfe of our botanical purfuits, we have long been in the habit of looking up to the learned Prefident of the Linnean Society, and have ever been happy to fhelter our-felves under his authority; it is not without reluƈance there-fore, that, in this inftance, we think ourfelves obliged to forfake his banners, and adhere to thofe principles by which we have been hitherto governed, in our choice of names.

LINNÆUS

LINNÆUS had followed preceding botanifts in uniting the plant, with the figure of which we now prefent our readers, to the genus Nymphæa, calling it NYMPHÆA *Nelumbo;* but on account of the very remarkable difference in the ftruċture of the fruit, later botanifts found it neceffary to raife it into a new genus: accordingly ADANSON, GÆRTNER, JUSSIEU, and WILLDENOW adopted LINNÆUS's trivial name, the vulgar appellation of the plant in the ifland of Ceylon. For the fake of uniformity of language, JUSSIEU gave it a Latin termination, and NELUMBIUM has been fince generally admitted. But Dr. SMITH, departing from his great mafter's precept, that a *fuitable* name is not to be changed even for a *better,* prefers CYAMUS, a name under which the fame plant is defcribed by THEOPHRASTUS, and therefore, doubtlefs, unobjeċtionable, had it been at all neceffary to feek a new one. It may be re-marked, however, that this name is not given by THEOPHRAS-TUS exclufively to this plant, as the worthy Prefident feems to have imagined; it before belonged to a leguminous plant, probably, fome kind of bean, common in Greece, and was applied to the Nelumbium merely on account of the fimilarity of the feed, juft as our Englifh voyagers give the name of apples, pears, and goofeberries, to fuch tropical fruits as bear fome fort of refemblance to the produce of their own country, and precifely as HERODOTUS had long before, in defcribing the fame plant, called it a rofe-coloured Lily; on which account, by the bye, another botanift may think, that CRINUM has every right and title to be preferred, and thus names may be altered without end.*

Dr. SMITH accufes us, in common with other modern writers, of confounding the NYMPHÆA *Lotus* with this plant, but certainly without juft ground. Indeed, no botanift can poffibly have miftaken the one for the other, fince the publi-cations of GÆRTNER and JUSSIEU, however the mythological hiftory of thefe plants may have been occafionally mifapplied. If any difficulties remained, thefe have been fince cleared up.

---

* It may appear, at firft fight, that as the chapter begins "῾Ο δὲ κύαμος φύεται," &c. that this name is applied exclufively to the plant of which THEOPHRASTUS is here fpeaking, but in fome copies the reading is "῾Ο δὲ κύαμος ἐν Ἀιγύπτῳ φύεται"; there is no occafion, however, to have recourfe to this reading, for the fentence is evidently continued from the end of the former chapter, where the author is treating of aquatic plants growing in Egypt, and confequently the repetition of the epithet Egyptian was unneceffary. In other places THEOPHRASTUS has himfelf ufed the word κύαμος to denote fome kind of pulfe, and it occurs repeatedly in the works of HIPPOCRATES with the fame meaning,

by the masterly papers of Savigny and De Lile, published in the first volume of the *Annales du Muséum d'Histoire Naturelle*, in 1802 (vide *Annals of Botany, vol. 2. page* 174.) which contain a very fair and satisfactory account of the Egyptian Lotus, and a careful distinction of it from the Egyptian Bean; together with every thing that has been said of these plants by Herodotus, Theophrastus, and other ancient writers.

Whether in the Egyptian mythology the Nymphæa Lotus became important only as a substitute for the Sacred Bean, as Dr. Smith presumes, we leave to others to decide; to us, however, it seems probable that a plant which made its appearance only with the inundation of the Nile, the source of all fertility in Egypt, and disappeared as the water receded, lying concealed in the arid sand, until revivified by the succeeding inundation, could not fail to be celebrated by the ancient Egyptians, independent of any resemblance it might bear to the Sacred Bean. Even the present inhabitants distinguish it by the name of the spouse of the Nile, a term expressive of its being considered by them as the symbol of the fertility about to be renewed by the sojourn of the waters upon the earth. Our present plant which grew, as Theophrastus informs us, in stagnant waters, and not in the lands overflowed by the Nile, appears to have had a less powerful claim to the adoration of the superstitious Egyptians.

Besides, the more frequent occurrence both of the flower and fruit of the Nymphæa *Lotus*, than of the Nelumbium, on the sculptured monuments and symbolic tables of the ancient temples of Egypt, militates against the President's opinion; as does also the blending of the fruit of the former plant with the ears of corn, to form the insignia of Isis, as the symbols of fertility and abundance, and the probable conjecture, that the Poppy was dedicated to Ceres, whose attributes are so similar to those of the Egyptian goddess, entirely on account of its resemblance to the *Lotus*. It is remarkable too, that if the Nelumbium was really the celebrated Lotus of antiquity Herodotus and Theophrastus, who have described both plants, should have agreed in applying the name of Lotus to the Nymphæa.

The conjecture of our learned friend, that the Beans, said to be forbidden by Pythagoras to be eaten by his disciples, were the fruit of the Nelumbium, although favoured by the circumstance, that this celebrated philosopher is supposed to have imbibed his doctrines from the Egyptian priests, may, perhaps, be rendered dubious by the apparent absurdity of proscribing the use of a vegetable altogether unknown in Greece; a conduct

duct much the fame as if an Englifh enthufiaft fhould preach up in London the neceffity of a total abftinence from yams !

The Nelumbium is no longer found in Egypt, but is common in moft parts of the Eaft-Indies, and appears to be held in high eftimation in China, where there are feveral varieties, if not diftinct fpecies. Is faid to occur likewife in the Weft-Indies, but it appears to us probable that this is a different fpecies.

The feeds of this plant preferve their vegetative properties for very many years; which makes it the more furprifing, that fuch a very ornamental and fragrant flower fhould not more frequently occur in our ftoves; but its proper culture does not feem to be as yet well underftood. It requires a deep ciftern with a confiderable depth of mud for its roots. The ancient Egyptians planted the feeds in balls of mud or clay, mixed with chaff, and thus funk them in the water; perhaps this practice might be fuccefsfully imitated.

Although feldom reared to perfection in this country, it bears the fevere cold of Pekin with impunity. Probably, if attention were paid to obtain feeds from the coldeft climes in which it is found, we might be more fuccefsful in cultivating it, with little or no artificial heat; at prefent, we believe, it has not with us been made to flower out of the ftove.

Our drawing was firft fketched from a very fine plant in bloffom at Mr. LIPTRAP's, at Mile-End, in the year 1797, and finifhed from one that made a moft magnificent appearance in the ftove of the Right Honourable CHARLES GREVILLE, at Paddington, in the fummer of 1804. Introduced in 1784, by the Right Hon. Sir JOSEPH BANKS, Bart.

A. reprefents a flower of the natural fize over a fmall leaf.

B. a diminifhed figure of the plant, fhewing the mode of its growth.

Syd Edwds del Pub by T Curtis, St Geo Crescent Feb 1 1806 F Sansom sculp

Cal. 5-fidus campanulatus. *Petala* 5, calyci inferta. *Bacca* 5-locularis, calyce obvoluta.

Specific Character and Synonyms.

MELASTOMA *corymbofa;* foliis feptemnerviis cordato-ovatis acutis ferratis nudis, corymbo paniculato terminali.

MELASTOMA 5. *Afz. in Herb. Banks?*

●

This very handfome fpecies of Melaftoma is a native of Sierra Leone, on the Weft coaft of Africa. We believe it was introduced into this country by our friend, Profeffor A f z e l i u s, and is hitherto undefcribed.

Our memoranda have been miflaid, but, if we miftake not, it belongs to the octandrous divifion, having eight ftamens, three of which are fterile. Requires to be kept in the ftove, but is not fo impatient of cold as might be fufpected from the place of its natural growth. Is propagated by cuttings. Our plant was received from Mr. Loddiges, Hackney.

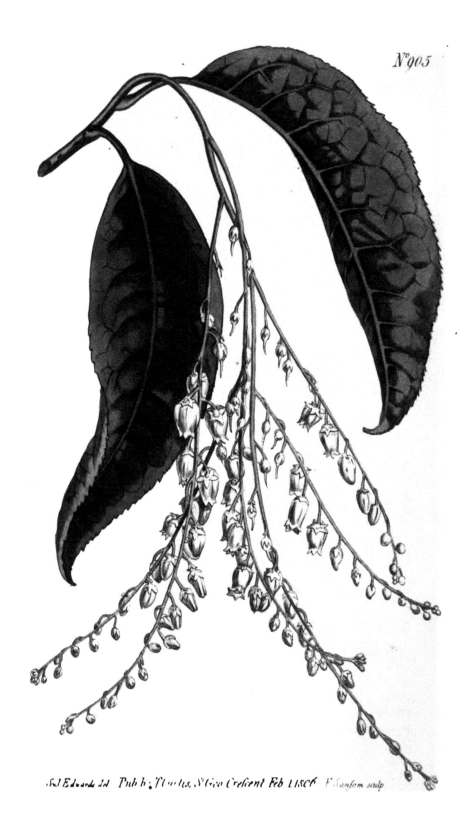

Syd Edwards del Pub h. J. Curtis St Geo Crefient Feb 1 1806 F Sanfom sculp

ANDROMEDA ARBOREA. TREE ANDROMEDA, or SORREL-TREE.

Clafs and Order.

DECANDRIA MONOGYNIA.

Generic Character.

Cal. 5-partitus. *Cor.* ovata: ore 5-fido. *Capf.* 5-locularis.

Specific Character and Synonyms.

ANDROMEDA *arborea ;* paniculis terminalibus, corollis fubpubefcentibus, foliis ellipticis acuminatis denticulatis. *L'Herit. Stirp. Nov. v.* 2. *Hort. Kew.* 2. *p.* 69. *Willd. Sp. Pl.* 565. *Mart. Mill. Dict. a.* 13.

ANDROMEDA *arborea ;* racemis fecundis nudis, corollis rotundo-ovatis. *Sp. Pl.* 565.

ANDROMEDA *arborea ;* ramis teretibus; foliis majufculis oblongo-ovalibus, acutiffime acuminatis, argute ferrulatis : panicula terminali, poly-ftachia : corollis pubefcentibus, ovoideo-cylindraceis : antheris linearibus, muticis. *Michaux Flor. Bor-Am.* 1. *p.* 255.

THE SORREL-TREE. *Catefb. Car.* 1. *p.* 71. *t.* 71.

Said to grow in its native foil, in the Alhegany-Mountains, into a tree fifty or fixty feet high. The fpecimen from which our drawing was taken forms a very large fhrub in Meffrs. WHITLEY and BRAME's Nurfery, in Old Brompton; the branches are pendent, and the long racemes of white flowers grow from their extremities. Bloffoms from July to September, and during this feafon in particular makes a very handfome appearance.

Syd Edwards del Pub. by T.Curtis St Geo. Crescent Feb.1.1806   F.Sansom sculp

ZIZIPHORA SERPYLLACEA. SWEET-SCENTED
• ZIZIPHORA.

****

Clafs and Order.

DIANDRIA MONOGYNIA.

Generic Charaƈer.

Cor. ringens: labio fuperiore reflexo, integro. *Cal.* filiformis. *Sem.* 4.

Specific Charaƈer and Synonyms.

ZIZIPHORA *ferpyllacea ;* capitulis terminalibus ovalibus, foliis ovatis fubferratis : floralibus fubfimilibus integerrimis ciliatis.

ZIZIPHORA *ferpyllacea ;* racemis terminalibus capitatis, foliis ovato-lanceolatis fubferratis : floralibus confimilibus. *Marfchall. v. Beberftein Terek u Kur, p.* 127. *Annals of Botany, v.* 2. *inedit.*

ZIZIPHORA *odoratiffima. Loddiges.*

Seeds of this alpine plant were received from Mount Caucafus, by Mr. LODDIGES, who kindly fent it us in flower in July laft. It continues feveral weeks in bloffom, and is at the fame time agreeable to the eye and grateful to the fmell.

Having compared our plant with the fpecimens fent from Caucafus by Mr. ADAMS to Sir JOSEPH BANKS, under the name we have adopted, we are certain of the identity of their fpecies. In thefe fpecimens as well as in ours the braƈes are not exaƈly fimilar to the leaves, being fmaller, rounder, and more acuminate, without notches, ciliated at the edge, and more ftrongly nerved. The filaments in this fpecies are extremely fhort, the anthers being nearly feffile in the faux of the corolla. In the fame colleƈion is another plant very nearly refembling this, except that the calyx is thickly covered with ftiff hairs; as far as we can judge in the dried ftate it appears to be a mere variety of this; Mr. ADAMS however confiders it as a diftinƈ fpecies, and calls it ZIZIPHORA *Poufchkini.* We have preferved as a fynonym the name by which Mr. LODDIGES received it, as we have fometimes found that the plants of thefe very diftant countries have been publifhed under thefe names long before we have known of it.

Syd Edwards del. Pub by T Curtis, St Geo Crefcent Feb 1 1806   F Sanfom sculp

CORONILLA CORONATA. CROWN-FLOWERED CORONILLA.

✱✱✱✱✱✱✱✱✱✱✱✱✱✱✱

Class and Order.

DIADELPHIA DECANDRIA.

Generic Character.

Cal. 2-labiatus : ⁴⁄₂ : dentibus superioribus connatis. *Vexillum* vix alis longius. *Legumen* isthmis interceptum.

Specific Character and Synonyms.

CORONILLA *coronata;* caulibus erectis flexuosis, foliolis novenis ellipticis: internis cauli approximatis, leguminibus pendulis.

CORONILLA *coronata ;* suffruticosa, foliolis novenis ellipticis : internis cauli approximatis, stipula oppositifolia bipartita. *Willd. Sp. Pl.* 3. *p.* 1151.

CORONILLA *coronata. Spec. Pl.* 1047. *Reich.* 3. 492. *Syst. Veg.* 669. *Jacq. Austr. t.* 95. *Hoffm. Germ.* 260. *Roth. Germ.* I. 318. II. 212.

CORONILLA *montana. Scop. Carn.* 912. *t.* 44. *Riv. tetr. t.* 93.

CORONILLA caule erecto, foliis undenis ovatis, floribus umbellatis, siliquis articulatis pendulis. *Hall. Helv.* 388.

COLUTEA siliquosa minor coronata. *Bauh. Pin.* 397. *Raii Hist.* 924.

COLUTEA scorpioides altera. *Cluf. Pan.* 46. *Cluf. Hist.* 1. *p.* 98.

COLUTEA scorpioides montana Clusii. *Ger. Emac.* 1300.

The stipulæ in this plant, if any, are so deciduous, that, when the plant is in flower, they are not seen, on which account we have omitted them in the specific character and added the more distinguishing mark of the pendulous seed-pods. It is properly an herbaceous perennial, for, although the lower part of the stem becomes woody, it perishes to the ground every year ; is a much handsomer growing plant than the Coronilla varia; the root survives our common winters in the open ground, unless the soil be too wet. Is a native of the southern mountains of Europe. Propagated by seeds. Flowers in the Summer months. Introduced in 1776 by Professor JACQUIN. Our drawing was taken at Mr. SALISBURY's Botanic Garden, Brompton.

Syd.t Edwards del   Pub by T.Curtis St. Geo. Crescent Feb.1 1806   F.Sansom sculp.

Generic Character.

Cal. 1-phyllus, ventricofus. *Petala* 5, unguiculata. *Capf.* fupera, femitrilocularis, apice dehifcens, polyfperma. SMITH.

Specific Character and Synonyms.

SILENE *fimbriata ;* floribus dichotome paniculatis, petalis femibifidis incifo-fimbriatis, corona bipartita, caly-cibus inflatis venofis pubefcentibus.

CUCUBALUS *multifidus. Adams in Herb. Bankf.*

LYCHNIS Behen alba fimilis, major perfoliata. *Buxb. Cent.* 3. *p.* 31. *t.* 57. ?

DESC. *Stalks* erect, two feet high, hairy, round. *Leaves* broad, heart-fhaped, quite entire, rugofe underneath, hifpid. on both fides : margin undulated, upper ones feffile, lower petiolated. *Petioles* winged, connate. *Inflorefcence* a dicho-tomous panicle. *Calyx* inflated, covered with a foft pubef-cence. *Corolla* five-petaled : claws of the petals very narrow and diftant, expanding into wings at the upper part ; limb divided half-way : lobes finely cut. Crown a bipartite procefs arifing from the middle of the limb, not its bafe. *Stamens* ten ; filaments the length of the petals, inferted into the elevated receptacle below the ovary ; anthers ovate greenifh. *Ovary* nearly globular, fomewhat flattened, half three-celled ; ftyles three ; ftigmas acute. *Capfule* one-celled, the diffepiments which extended from the circumference half-way to the recep-tacle in the ovary now difappearing. *Receptacle* of the feeds conical, central, unconnected. *Seeds* reniform rugofe.

This

This has altogether the habit of Cucubalus Behen of LIN-NÆUS (SILENE *inflata* of SMITH) of which it is doubtlefs a congener. However averfe from unneceffarily changing names, we entirely agree with our friend Dr. SMITH in the propriety of feparating thefe plants from CUCUBALUS *bacciferus* and uniting them with the genus Silene, of which the prefent fpe-cies has altogether the charaƈter; nor is the Behen perfeƈtly free at all times from thefe proceffes, which forms what LINNÆUS calls the *corona*, as is obferved by Dr. SMITH, and before him by that accurate Botanift POLLICH.

Native of Mount Caucafus; perfeƈtly hardy; propagated by feeds, which it produces plentifully. Introduced by Mr. LODDIGES, from whom we received it in flower in July laft, under the name of CUCUBALUS *fimbriatus*.

Syd Edwards del.   Pub. by T Curtis, St Geo Crescent Mar 1.1806   F.

Generic Character.

Legumen compreffum, cochleatum. *Carina* corollæ a vexillo defleɛtens.

Specific Character and Synonyms.

MEDICAGO *carfienfis ;* pedunculis multifloris, leguminibus cochlcatis utrinque compreffis, aculeis fubulatis reɛis, ftipulis dentatis, caule ereɛo. *Willd. Sp. Pl.* 3. *p.* 1412.

MEDICAGO *carfienfis ;* caule ereɛo, radice perenni reptante, floribus umbellatis, leguminibus cochleatis fetofis. *Jacq. Coll.* 1. *p.* 86. *Ic. rar.* 1. *t.* 156. *Hoft Syn.* 417.

MEDICA cochleata polycarpos, capfula fpinofa minore, perennis ciliaris five capfulis ciliaribus nigris. *Morif. Hift.* 2. *p.* 154.

MEDICA ciliaris Guilandini. *Raii Hift.* 965.

This fpecies of Medicago is undoubtedly diftinɛ from all the fuppofed varieties of MEDICAGO *polymorpha*, particularly in having a perennial creeping root and upright, fquare, almoft fhrubby ftalks.

Found by BURSATI in the Çarftian mountains in Carniola, by whom feeds were fent to Profeffor JACQUIN. Although feemingly confidered by him as entirely new, it was known to fome of the older Botanifts ; RAY gathered his plant in the mountains of Carinthia. REICHARD, in his edition of the Species Plantarum, added the Synonyms of RAY and MORISON to the *ciliaris*, in which he was followed by Profeffor

·MARTYN,

Martyn, in his edition of Miller's Dictionary; but Willdenow is certainly right in having applied them to this plant.

Flowers in June and July; is a hardy perennial, but like other alpine plants, apt to perish in our humid winters.

Our drawing was taken from a plant sent us by Mr. Loddiges. We had also a specimen some years before from the Botanic Garden at Brompton. We learn from Mr. Donn, in his Hortus Cantabrigiensis, that it was introduced in 1790, probably by himself.

Syd. Edwards del Pub. by T. Curtis, S.t Geo Crescent Mar 1. 1806

HALESIA TETRAPTERA. FOUR-WINGED SNOW-DROP-TREE.

****

Clafs and Order.

DODECANDRIA MONOGYNIA.

Generic Charaƈer.

Cal. 4-dentatus, fuperus. *Cor.* 4-fida. *Nux* 4-angularis, 4-locularis. *Sem.* folitaria.

Specific Charaƈer and Synonyms.

HALESIA *tetraptera ;* foliis ovatis acuminatis, venis fubtus pilofis, alis fruƈus æqualibus. *Willd. Arb.* 137. *Ejufd. Sp. Pl.* 2. *p.* 849.

HALESIA *tetraptera. Sp. Pl.* 636. *Reich.* 2. *p.* 417. *Michaux Fl. Bor-Amer.* 2. *p.* 40. *Gært. Fruƈ.* 1. *p.* 160. *t.* 32. *Cavan. Diff.* 6. *p.* 338. *t.* 186. *Ellis in Aƈ. Ang. v.* 51. *p.* 831. *t.* 22. *f. A. Mart. Mill. Diƈ. a.* 1.

FRUTEX padi foliis ferratis, floribus monopetalis albis campaniformibus, fruƈu craffo tetragono. *Catefb. Car.* 1. *p.* 64. *t.* 64.

This fine fhrub was named in honour of the learned and venerable STEPHEN HALES, D. D. F. R. S. by JOHN ELLIS, Efq. who firft raifed it in this country from feeds fent over by Dr. ALEXANDER GARDEN, in 1756. It is a native of South-Carolina, where it grows by the fides of rivulets fhaded by wood. Is perfeƈly hardy. Flowers in April and May ; but not with us, as in America, entirely before the appearance of the leaves.

Propagated by feeds, and as thefe, in favourable feafons, are not unfrequently perfeƈed here, we are furprifed that a fhrub of fo much beauty fhould not be more generally met with in our pleafure-grounds. According to Dr. GARDEN, the fruit is alfo very pleafant to the tafte.

G.d Edwards del Pub by T. Curtis, N.   Crefcent Mar. 1 1800 F Sanjom sculp

CYPRIPEDIUM PARVIFLORUM. YELLOW LADIES SLIPPER.

✶✶✶✶✶✶✶✶✶✶✶✶✶✶✶✶✶✶

Clafs and Order.

GYNANDRIA DIANDRIA.

Generic Character.

Petala 4, cruciatim difpofita. *Nectarii* labium inferius ventricofum, inflatum, calceiforme.

Specific Character and Synonyms.

CYPRIPEDIUM *parviflorum ;* petalis lateralibus linearibus contortis calceolo avenio longioribus.

CYPRIPEDIUM *parviflorum ;* lobo ftyli fagittæformi bafi deflexo, labello petalis breviore compreffo. *Swartz Orchidæ. Tracts on Botany, p.* 207. *Salifbury in Linn. Tranf. v.* 1. *p.* 77. *t.* 2. *f.* 2.

CYPRIPEDIUM *Calceolus ;* minutim pubefcens : caule foliofo : laciniis calycis exterioribus oblongo-ovalibus, acuminatis ; interioribus lineari-bus *confertifque :* calceolo luteo. *Michaux Fl. Bor-Amer. v.* 2. *p.* 161.

HELLEBORINE calceolus dicta, mariana, caule foliofo, flore luteo minore. *Pluk. Mant. p.* 101. *t.* 488. *f.* 2.

This fpecies of Ladies-Slipper is an inhabitant of North-America, from New-England to North-Carolina. It comes very near to the European fpecies, and we fuppofe has been miftaken for the fame by MICHAUX, in whofe fpecific defcription *confertis* feems to be an error of the prefs for *contortis.* It is a taller plant, more pubefcent; lateral or interior petals longer, narrower, and more curled; and the nectarium or flipper is of a plain yellow colour without veins.

Our

Our drawing was taken at Mr. Woodford's, at his late refidence at Vauxhall, early in June.

Jussieu calls fegments of the calyx, both the petals and nectarium, as we, in conformity to Linnæus and moft Botanifts, call them; Swartz confiders our petals as calyx and the nectarium as corolla. Dr. Smith has very properly remarked that there is no end to difputes of this kind; to us the language of Linnæus appears the moft eafily intelligible, and the name of nectarium, fo offenfive to many modern Botanifts, is conveniently applied to thofe parts of a flower that have an anomalous form, although they may not always be organs for fecreting or retaining the honey. To avoid the impropriety of borrowing a name from a function, which the part does not perform, fome botanifts have propofed to fubftitute *parapetalum* for *nectarium*; but it feems hardly worth while to change an eftablifhed word, well underftood; and indeed we believe that the cafes, where thefe parts are not really receptacles of honey or fome analogous fluid, are fewer than is generally fuppofed.

C. Edwards ad. Pub by T. Curtis, St. Geo. Crescent Mar. 1 1806. F. Sansom sc.

CAMPANULA MACROPHYLLA. LARGE-LEAVED BELL-FLOWER.

Claſs and Order.

PENTANDRIA MONOGYNIA.

Generic Charaſter.

Cor. campanulata, fundo clauſo valvis ſtaminiferis. *Stigma* trifidum. *Capſ.* infera, poris lateralibus dehiſcens.

Specific Charaſter and Synonyms.

CAMPANULA *macrophylla;* capſulis obteſtis trilocularibus, foliis cordatis crenatis ſubtus tomentoſis, floribus in racemo compoſito nutantibus ſecundis.

CAMPANULA alliariæfolia. *Adams ?*

DESCR. *Root* biennial ? tap-ſhaped. *Stalk* round, ereſt, ſtriated, woolly, and hairy, branched at the top only and ſometimes terminated in a ſimple raceme. *Radical leaves* large, heart-ſhaped (in our ſpecimen haſtate, but this form does not appear to be conſtant) crenate, tomentoſe underneath, *cauline* on footſtalks gradually ſhortening upwards : *floral* quite ſeſſile. *Flowers* white, nodding, ſecund, at the ends of the branches, on ſhort curved footſtalks, which grow ſingly from the axils of the floral leaves. *Calyx* of five lanceolate entire ſegments, with their margins rolled back, ereſt, with the angles lengthened, reflected upon and adhering to the *germen,* which is top-ſhaped, irregularly ſulcated, three-celled, and terminated by a trifid, finally revolute *ſtigma. Corolla* bell-ſhaped : tube three times longer than calyx, widening gradually upwards : limb five-cleft, ſegments patent, ciliated at the edges ; angles between the ſegments as if pinched into a tooth-
like

like procefs, vifible before the bloſſom is expanded. *Valves* fupporting the ſtamens ovate, ciliated : *filaments* ſhort, capillary : *anthers* linear. The whole plant abounds with a clammy greeniſh milk.

For this new ſpecies of Campanula we are indebted to Mr. LODDIGES, the product of ſeeds ſent him from Mount Caucaſus. Specimens of the ſame plant are in the collection tranſmitted from that country to Sir JOSEPH BANKS, by Count MOUSHIN POUSHKIN ; but the ticket containing the name given by Dr. ADAMS having been loſt, we are not certain that we apply from his liſt the one intended, and the leaves not appearing to us to bear a good refemblance to thoſe of Alliaria, we have preferred that by which we ſaw the ſame plant defignated in Mr. VERE's fine collection at Kenſington-Gore.

Is perfectly hardy. Flowers in July and Auguſt. Propagated by feeds. Being of large growth it requires room, and is, on that account, more adapted to ornament extenſive plantations, than the confined parterre.

Syd. Edwards del. Pub. by T. Curtis, St Geo Crescent Mar 1 1806. F. Sansom sculp.

EUCOMIS PUNCTATA. SPOTTED-LEAVED EUCOMIS.

****

Class and Order.

HEXANDRIA MONOGYNIA.

Generic Character.

Cor. infera, fexdivifa, perfiftens; laciniis rectioribus. *Stam.* ferto membranaceo brevi corollæ adnato connexa. *Germ.* trigonum, membranam ftaminilegam longe fuperans. *G.*

FRITILLARIA. *Linn.* BASSILÆA. *Juff.*

OBS. MASSONIÆ *confinis; dignofcenda tamen tubo vix ullo, germine fupra fertum ftaminilegum longe exftante, corollæ laciniis neve retroflexis vel etiam recurvatis; exque habitu, hic enim educitur* MASSONIÆ *umbella vel corymbus in racemum fpicatum coma variæ magnitudinis terminatum. Graviffime fallimur quando ad* Nᵐ· 840 *hujus generis radicem dicimus " Bulbum fquamofo-tunicatum ut in* LILIO;" *eft enim " Bulbus tunicatus, tunicis craffioribus," parum quidem ab illo* MASSONIÆ *recedens. G.*

Specific Character and Synonyms.

EUCOMIS *punctata* foliis pluribus, oblongo-lanceolatis, canaliculato-depreffis; racemo elongato-cylindraceo; coma brevi microphylla; bracteis inclufis pedicellos fubæquantibus; corolla rotata; ftaminibus divergenter patentibus. *G.*

EUCOMIS *punctata. Hort. Kew.* 1. 433. *L'Herit. Sert. Angl.* 18. *t.* 18. *Mart. Mill. Dict.* 4. *Willd. Sp. Pl.* 2. 93.

ORNITHOGALUM *punctatum. Thunb. Prod.* 62.

ASPHODELUS *comofus. Houtt. Linn. Pfl. Syft.* 11. *p.* 381. *t.* 83.

The trivial name is taken from the curious dotting of the ftem and leaves.—Introduced by Mr. JOHN GRÆFER, in 1783, from the Cape of Good Hope. Flowers in July. Its fcent feems

to

to us not unpleasant. The leaves do not lie flat on the ground as those of many of the species do; but are upright and divaricately patent; pedicles little longer than the corolla, about equal to the bractes, which are somewhat coloured and concave; corolla stellately patent and parted almost to the base; filaments shorter than corolla, connate, but only for a very little distance, divergent, somewhat incurved; both filaments and corolla are at first white, but turn green in time. Germen ovate-fastigiate; style curved. A common greenhouse plant, of most easy culture.

Our drawing was made from a specimen sent by Mr. Bucha-nan, Nurseryman, at Camberwell. *G.*

Syd Edwards del   Pub by T Curtis St Geo Crescent Mar 1 1806   F Sansom

PHALANGIUM LILIAGO (β). LESSER GRASS-LEAVED PHALANGIUM.

****

Clafs and Order.

HEXANDRIA MONOGYNIA.

Generic Character.

Cor. infera, hexapetalo-partita, tota patens vel a bafi connivens, perfiftens. *Filam.* filiformia, nuda. *Stylus* affurgens, furfum incrafcefcens. *Stigm.* hianter obtufum. *G.*

OBS. Radix perennis, fibrofa; fibris fimplicibus, carnofis, craffiufculis, fafciculatis. Folia radicalia, ex linearibus longe attenuata, canaliculata. Flores albi, faepius cum virore. Caulis annuus, erectus, fimpliciffimus vel ramofus; pedunculi bracteati, uniflori, obfcurius uniarticulati, fpicatim digefti. Differt ASPHODELO filamentis bafi haud fornicatim dilatatis atque germini coaptatis; capfulae quoque fubftantia, quae tenuior et abfque omni parenchymate; ANTHERICO filamentis nudis; fed praeprimis habitu.—Iftuc fubjicienda ANTHERICUM Liliaftrum fupra N^{um.} 318 et ANTHERICUM ramofum. G.

Specific Character and Synonyms.

PHALANGIUM *Liliago* caule fimpliciffimo; foliis gramineo-anguftis, fubulato-linearibus; racemo multifloro rariufculo, fpicato; corolla ftellato-patente, laciniis oblongo-lanceolatis. *G.*

PHALANGIUM *Liliago. Schreb. Spicil.* 36.

ANTHERICUM *Liliago. Linn. Sp. Pl.* 445. *Suec. n.* 290. *Jacq. Hort.* 1. *t.* 82. *Pollich. Pal. n.* 335. *Krock. Siles.* 528. *Vill. Dauph.* 2. 267. *Hort. Kew.* 1. 449. *Flor. Dan. t.* 616. *Hoffm. Germ.* 121. *Gaertn. Sem. et Fr.* 55. *t.* 16. *f.* 1. *Flor. Arragon.* 44. *Willd. Sp. Pl.* 2. 141. *Desf. Flor. Atl.* 1. 304.

PHALANGIUM, &c. *Hall. Helv. n.* 1207.

ANTHERICUM caulibus non ramofis. *Guett. Stamp.* 1. 128.

PHALANGIUM

PHALANGIUM parvo flore non ramofum. *Baub. Pin.* 29.
     Mor. Hift. f. 4. *t.* 1. *f.* 10. *Park. Parad.*
     150. 3. *t.* 151. *f.* 2. *Ger. Herb.* 44. *f.* 2.
     J. Baub. Hift. 2. 635. *cum Ic.*

LILIAGO Cordi. *Lob. Ic. p.* 48. *R.*

LILIAGO. *Cordi Hift.* 2. *c.* 106. *p.* 190. *b.*

(α) major. G.

(β) minor. G.

The prefent plant, with the two others mentioned in our obfervation above, are too diftinct from thofe plants with which they have been ufually arranged, under the generic title of ANTHERICUM, to be any longer permitted to remain in the fame fection; we have accordingly feparated them under the name already adopted by JUSSIEU, as well as fome of the older Botanifts.

The reafons alleged by JACQUIN for fuppofing HALLER's plant to be a different fpecies from this, we think ill founded; he appears to us to have miftaken differences of terms for differences of things; "the petiolated petals" of HALLER being no other than the claws or narrowed bafe of the fegments of the corolla; nor are "the petals of two orders" of the fame any thing more than the inner and outer fegments of the flower.

A hardy common plant; native of Algiers, Spain, France, Switzerland, Italy, and Denmark. The variety (α) is handfomer than the prefent. Differs from P. *Liliaftrum* by its ftellately patent corolla, and from *ramofum* by its fimple ftem. G.

Syd. Edwards del   Pub by T. Curtis S.ᵗ Geo. Crefcent Mar 1 1806 F. Sanfom

Specific Character and Synonyms.

AMARYLLIS *revoluta* foliis anguſtis, lorato-linearibus, cana-
      liculatis ; umbella pluriflora ; corolla anguſtius
      infundibuliformi, extrorſum curvata ; laciniis
      longinque recurvato-patentibus ; tubo obſolete
      angulato iſtis 2–3-plo breviore germine 2–3-
      plo longiore ; ſtigmate hiante.   *G.*
AMARYLLIS *revoluta.   Hort. Kew.* 1. 419.   *L'Herit. Sert.*
      Angl. 14.   *Mart. Mill. Dict.* 14.   *Willd. Sp.*
      Pl. 2. 57.

    This handſome plant is a native of the Cape of Good Hope ;
and was originally introduced into the Kew-Gardens by
Mr. MASSON, in 1774 ; but the ſpecimen, from which our
drawing has been made, was imported by Mr. HIBBERT, with
whom it flowered five or ſix years ago ; the leaves were de-
cayed when ſeen by our draughtſman, and we fear the bulb has
ſince ſhared their fate, as we have ſought for it in vain in the
ſame collection.
    Leaves narrow, quite linear ; ſtalk purple-brown ; um-
bel (in thoſe plants which have bloomed in this country) four
to ſix flowered ; corolla narrow-turbinate, recurvedly patent
full as far as the middle, white ſuffuſed with different ſhades
                                    of

of rofe-colour; pedicles round, black-purple, bent outwards, feveral times longer than the elliptic green germen, which is unufually fmall in proportion to the corolla; the fegments are without the undulate edge that we fee in the *vittata*; organs affurgently declinate; ftyle rofe-coloured. Sweet-fcented; blooms in September; needs nothing more than protection from froft, and, perhaps, will do at the foot of a fouthern wall, as well as moft of its Cape congeners. A very rare fpecies, and if Mr. HIBBERT's plant is really loft, we fufpect it is not now to be found in any European collection. We have reafon to think the bulb fufficiently diftinct from both *Belladonna* and *vittata*; but trufting to the feeing of it another year, we loft the opportunity of taking fo complete a defcription of the whole plant as we now wifh we had done. The leaves come very near to thofe of AMARYLLIS *Belladonna*, as the flowers do to thofe of AMARYLLIS *vittata*. G.

Syd Edwards del Pub by T Curtis; St Geo Crescent 1.1806 F Sansom sculp.

Uvularia Chinensis. Brown-Flowered Uvularia.

****

Class and Order.

Hexandria Monogynia.

Generic Character.

Cor. infera, fexpartita, campanulata, laciniis rectis; unguibus fovea nectarifera oblonga excavatis. *Stam.* (*fæpius*) brevif-fima. *Stigmata* 3, reflexa. *Capf.* trigona; *femina* fubrotunda comprefsa. Convallariæ habitu, Fritillariæ charactere confinis.

Specific Character and Synonyms.

UVULARIA *chinenfis* (*fimpliciter atque fubcorymbofe ramofa*) foliis ovato-lanceolatis, acuminatis, racemis faf-ciculatim 2—4 floris, fingulis fafciculis folio con-formi bracteatis; corolla cyathiformi-campanu-lata, angulofa, bafi calcarato-nodofa; filamentis antheris aliquoties longioribus. G.

This fingular as well as new fpecies is (as we learn from the Bankfian Herbarium) a native of China, and flowered two years ago in the Kew Gardens. Our drawing was made in September laft from a plant that bloomed in Mr. Hibbert's confervatory at Clapham.

The following defcription is taken from a recently dried fpecimen, in which however the flowers were fo far deftroyed by preffure that we could not make out either the form or even fcite of the nectary, nor difcover whether all or only the alter-nate fegments terminated in the fame kind of blunt fpur-like knob.

Stem herbaceous, about a foot and half high, angular, fub-geniculately flexuofe, diftantly leafy, branched upwards, branches fimple, corymbofely arranged, patent; leaves ovate-lanceolate, acuminate, fhortly petioled, nerved; thofe of the

ftem

ftem broader, elliptic, diftant; thofe of the branches narrower, farther acuminate, and more clofely fet together; racemes one to four-flowered, rameous, axillary to the leaves, pedicles fafciculate; peduncle fhorter than the fafcicle, which laft has a leafy bracte at their bafe of the fame form as the upper leaves, fo that where there is a raceme it appears as if there were two oppofite leaves; corollas cernuous, longer than pedicles, cupped-campanulate, brown without, knottedly angular at the bafe as if fhortly and bluntly fpurred; ftamens equal to corolla and piftil; filaments fubulate-linear, two to three times longer than the anthers; germen turbinately triquetral, feveral times fhorter than the ftyle, ftigmas patent, revolutely recurved. This fpecies differs from all its congeners yet known in the length of the filaments.   G.

J.v.d. Edwards del. Pub by T. Curtis, St Geo Crescent Apr 1 1806.   F. Sansom sculp

HYPOXIS SERRATA (β). LARGE-FLOWERED SERRATE-LEAVED HYPOXIS.

****

Clafs and Order.

HEXANDRIA MONOGYNIA.

Generic Character. *Vide* Nᵘᵐ· 709.

Specific Character and Synonyms.

HYPOXIS *ferrata.* *Vid. fupra in* Nº· 709.
(β) flore majore, miniato-aurantiaco; bracteis atque foliorum
ferratura fere obfoletis. G.
HYPOXIS *linearis.* *Bot. Rep. t.* 171. *fig. parum bona.*

This very handfome variety was imported by Mr. HIBBERT
fome years ago, from the Cape of Good-Hope, and agrees in
every refpect with the variety before defcribed in this work,
except in fize, colour, and obfoletenefs of the bractes, and
denticulated ferrature of the leaves. G.

AMARYLLIS REVOLUTA.

Since the publication of our account of this plant (Nº 915)
we are accidentally enabled to add the defcription of the bulb,
and correct that of the leaves there given.—The fpecific cha-.
racter fhould be thus amended.

AMARYLLIS *revoluta* foliis fcapum fubæquantibus ambienter
        fafciculatis, recumbenter effufis, a principio
        acuminato-attenuatis, canaliculato-depreffis,
        fubundatis; fcapo eccentrico; umbella pluri-
        flora; corollis angufte et longe infundibuli-
        formibus, curvato-nutantibus, laminis recurvo-
        patentibus, tubo 2-3-plo longioribus; ftigmate
        fubtrilobo hiante. G.
AMARYLLIS *variabilis.* *Jacq. Hort. Schœnb. v.* 4. icon,
        cujus adhucdum deeft defcriptio. " BAUER's
        " SKETCHES," *Fig.* abfque dato *nomine,* in
        Muf. Banks.

Bulb

Bulb largiſh, ovate-oblong, upwards lengthened, attenuated, ſquarroſe, partly above ground.—Leaves many, ſpringing from the crown of the bulb in a recumbently patent faſcicle, nar-row, long (one to two feet?), attenuated from their baſe, acuminate, cuſpidate, deeply channelled, ſubtriquetral; central very narrow, more erect; ſcape iſſuing on the outſide of the faſcicle. We were miſtaken in ſtating the foliage to be like that of A. *Belladonna,* it comes nearer, in fact, to that of A. *longifolia.* Pedicles often ſhorter than in our figure, and, as well as the ſcape, not always coloured. Segments of the corolla lanceolate, tube obſoletely trigonal; the reſt as in the deſcrip-tion and ſpecific character given in N° 915.

Differs from *longifolia,* to which it comes the neareſt, in the corolla being more narrowly funnel-form, in a proportionally ſhorter tube, and in having the laminæ further recurved. G.

ORNITHOGALUM SQUILLA (α). COMMON RED-ROOTED SEA-ONION, or OFFICINAL SQUILL.

****

Class and Order.

HEXANDRIA MONOGYNIA.

Generic Character.

Cor. infera, hexapetaloideo-partita, radiato-paffa, femel (in *nutante* nempe) connivens, bafi ftaminigera, perfiftens. *Fil.* latitudine varia, nuda, divergentia, alterna communiter latiora. *Stylus* fetiformis ftigmate inconfpicuo, vel brevis aut etiam fubnullus eodem capitato-trilobo. *Semina* fubglobofa, nunc angulatim vel femel paleaceo-preffa. *G.*

Obs. *Bulbus tunicatus. Folia radicalia, feriatim de anguftis et lineari-loratis lato-lanceolata, fucculenta, craffiufcula, teneriora. Flores nunquam cærulefcentes neve purpurafcentes, quo folo fufpicor figno* ORNITHOGALUM *dignofcendum a* SCILLA. *Scapus fimpliciffimus, teres, multiflorus, fpicatim vel corymbofe aut etiam thyrfoideo-racemofus; bracteæ membranaceæ. Capf. membranacea, oblongo- vel ovato- trigona, femel molendinacea (trialato-triloba) complanatis.* Confer OBS. *in* SCILLAM, *ad Num.* 746. *G.*

Specific Character and Synonyms.

ORNITHOGALUM *Squilla* bulbo maximo, globofo-ovato; imo rhizomate, umbonatim extanté; fummis tunicis fquarrofo-emicantibus; inflorefcentia folia lanceolata canaliculata diu præveniente; bracteis calcaratis? racemo cylindraceo, graciliori, confertiufculo, faftigiante; filamentis plano-fubulatis, corolla fatis brevioribus. *G.*

ORNITHOGALUM *maritimum. Tournef. Inft.* 381. *Brotero Flor. Lufit.* 1. 583. *Lamarck Flor. Fran.* 3. 276.

SCILLA *maritima. Syft. Veg.* 328. *Mat. Med.* 94. *Hort. Kew.* 1. 443. *Willd. Sp. Pl.* 2. 126. *Desf. Fl. Atl.* 1. 297. *Lil. à Redouté, t.* 116. *Lam. & Decand. Flor. Fran.* 3. 214. *Link & Hoffm. de* SCILLA. *Ann. Bot.* 1. 101. *Woodv. Med. Bot.* 322. *t.* 118.

SQUILLA. *Plantæ officinales, Amæn. Acad.* 4. 14.

SCILLA

SCILLA rufa magna vulgaris. *J. Bauh. Hift.* 2. 615. *Ic.*
SCILLA vulgaris radice rubra. *Bauh. Pin.* 73.
(α) radice rubra.
PANCRATIUM. *Cluf. Hifp.* 293. *Hift.* 171. cum herbæ
et bulbi icone abfque inflorefcentia.
(β) radice alba.
SCILLA hifpanica. *Cluf. Hifp.* 290, 291. *Hift.* 171. cum
iconibus.

This well known vegetable is a native of all the countries
bordering on the Mediterranean, as alfo of Brittany and Nor-
mandy; it has been found growing in the very fand of the fea-
fhore, and again, at the diftance of a hundred miles inland,
for inftance, at the foot of the Eftrella mountains; fo that, as
Link obferves, *maritimum* is rather a fallacious appellation.
By the Spaniards it is called *Cebolla albarrana*. The bulbs are
annually imported by our druggifts, for whofe purpofes both
varieties are ufed indifferently: they are efteemed powerfully
diuretic, and adminiftered chiefly in dropfical and afthmatical
cafes.

Blooms in July and Auguft, the leaves appearing in October
and November. MILLER fays the plant foon decays in our
gardens, and attributes the decline to want of fea-water, which
cannot, however, well be the caufe, as its natural fituation is
often at a great diftance from the fea, as we ftated above; with
us it has been preferved for thefe three years in vigour, planted
in a large garden pot and fheltered during winter in a common
garden frame; nor do we yet difcover the leaft fymptom of
decay. The root is frequently as big as a child's head, and often,
when frefh imported, throws out the flowering ftem while lying
in the fhop windows; the fpike is fometimes a foot or more
in length; pedicles rather fhort, filaments nearly equal; feed-
veffel alately three-lobed, a fhape that GÆRTNER terms *molen-
dinaceus*; feeds black, flat, chaff-like.

While SCILLA and ORNITHOGALUM continue to be kept
apart by the prefent barrier, which we think the only one there
is, we can have no doubt under which to range this fpecies.
BROTERO obferves, that when LÆFLING and, after him,
LINNÆUS, ftate ORNITHOGALUM *pyramidale* to be of Portuguefe origin, they have moft probably miftaken *maritimum* for
it, as *pyramidale* is certainly not a native of Portugal. G.

S.d Edwards del Pub by T Curtis, St Geo Crescent Apr 1 1806 F Sansom sculp

SCILLA AUTUMNALIS (α). PURPLE-FLOWERED AUTUMNAL SQUIL.

Clafs and Order.

HEXANDRIA MONOGYNIA.

Generic Charaƈer.

Cor. hexapetalo-partita laciniis radiato-paffis, aut fexfida iifdem campanulato-conniventibus, hinc, *filamentis* magis aut minus corolla concretis, HYACINTHUM contingens, inde, ipfis ufque ad bafin liberis, ORNITHOGALO confluens, cujus equidem cætera præter colorem qui hic nunquam exalbefcit nifi per varietatem. *Confer quæ diximus fupra, ut et ad Nᵘᵐ·* 746. *G.*

Specific Charaƈer and Synonyms.

SCILLA *autumnalis* foliis pluribus, angufto-loratis, obtufe canaliculatis; racemo corymbofe fpicato; pedicellis affurgentibus; braƈeis minutis, modo obfoletis; corolla nondum expanfa turbinata vertice fubtruncatim atque umbilicatim depreffa, aperta toto radiato-patente; germine tritorofo-globofo ftylo fetaceo 2-3-plo breviore. *G.*

SCILLA *autumnalis. Sp. Pl.* 443. *Cavan Ic.* 3. *p.* 38. *t.* 274. *Willd. Sp. Pl.* 2. 130. *Eng. Bot. t.* 78. *Curt. Flor. Lond.* 301. *Hort. Kew.* 1. 145. *Link & Hoffm. de Scilla apud Ann. of Bot.* 1. 106. *Desf. Fl. Atl.* 1. *p.* 301. *Brotero Flor. Lufit.* 1. 527. *Lam & Decand. Flor. Fran.* 3. 212.

SCILLA radice folida, foliis fetaceis, floribus faftigiatis pedunculatis arcuatis ex ala tuberculi mamillaris. *Guett. Stamp.* 1. *p.* 131. *Dalib. Paris.* 102.

HYACINTHUS autumnalis minor. *Raii Syn.* 373.

HYACINTHUS ftellaris autumnalis minor. *Baub. Pin.* 47.

HYACINTHUS autumnalis major et minor (*cum Iconibus bonis*). *Cluf. Hift.* 185.

(α) major; floribus purpurafcentibus vel cærulefcentibus.

(β) minor; floribus prædiƈis variantibus coloribus.

We

We fufpeᴄt our prefent variety to be of continental extrac-
tion, although the fpecies is a native of our ifland; growing in
feveral of the weftern diftriᴄts, and has even been found in
the neighbourhood of London. The Braᴄles (which are generally obfolete or fo inconfpicuous that they have been
overlooked and ftated not to exift by moft authors) were very
evident, though minute, in the prefent fpecimen, which
flowered in Mr. HIBBERT's garden at Clapham ; GUETTARD,
in his account of the plants growing about *Eftampes,* is the only
author we remember who mentions their prefence ; he terms
them *tuberculi mamillares.* DESFONTAINES found this fpecies
on the coaft of Africa with blue flowers ; as did LINK and
HOFFMANSEGG in Portugal, where, as they ftate, its inflorefcence precedes foliation ; a circumftance alfo obferved by
Dr. SIMS in this country, who fuppofes that when the leaves
accompany the flowers, it is the lefs natural mode, occafioned
by a particular wet feafon ; blooms from Auguft to September.

The Portuguefe Profeffor of Botany, BROTERO, obferves
that the variety which he found in the province of Eftrema-
dura was twice the fize of that which grew in the province of
Beira ; he does not notice the peculiarity in the flowering
mentioned by LINK and his fellow traveller.

The leaves grow on through the winter, dying away in
the fpring, after the manner of thofe of ORNITHOGALUM
Squilla. G.

S. Edwards del Pub by T Curtis, St Geo Crescent Apr 1 1806   F Sansom sculp

KÆMPFERIA ROTUNDA. ROUND-ROOTED GALANGALE.

Clafs and Order.

MONANDRIA MONOGYNIA.

Generic Chara&er.

Cal. obfoletus. *Cor.* 6-partita : laciniis tribus majoribus patulis, unica bipartita. *Stigma* 2-lamellatum.

Specific Chara&er and Synonyms.

KÆMPFERIA *rotunda ;* foliis lanceolatis petiolatis. *Flor.*
    Zeyl. 9. *Mat. Med. p.* 35. *Sp. Pl.* 3. *Willd.*
    15. *Reich.* 5. *Woodville Med. Bot.* 361. *t.*
    133. *Roxb. Corom. Pl.*
KÆMPFERIA *longa. Jacq. Hort. Schoenb.* 317. *Redouté*
    Lil. 49.
ZEDOARIA rotunda. *Baub. Pin.* 36. *Raii Hift.* 1340.
    Blackwell, t. 399.
MALANKUA. *Rheed. Mal.* 11. *p.* 17. *t.* 9.

We have no doubt but that this plant is the KÆMPFERIA
rotunda of LINNÆUS, and as little that it is the fame as is
figured by JACQUIN in his magnificent work the Hortus
Schoenbrunnenfis, and fince by REDOUTE in his Liliacées under
the name of KÆMPFERIA *longa.* Upon what grounds JACQUIN
confidered it as a different fpecies we can form no conjefture.
    The flowers appear early in the fpring, fome time before
the leaves, and have a very pleafing fcent, efpecially as they
dry. They grow immediately from the root, feveral in fuc-
ceffion, but feldom more than one or two are open at the fame
time. The organs of fruétification are very fimilar to thofe

of AMOMUM *exſcapum*, as figured in the Annals of Botany, vol. 1. pl. 13.

The college of phyſicians, both of London and Edinburgh, upon the authority of LINNÆUS, have referred the Zedoary of the ſhops to this plant; but the form of the roots as they occur at our druggiſts correſponds much better with thoſe of AMOMUM *Zerumbct*. But the roots of the larger Galangale, which LOUREIRO ſuppoſes to be ſold indiſcriminately for Zedoary or Galangale, are, at leaſt as they occur in our ſhops, totally different. Dr. ROXBURGH thinks that his CURCUMA *Zedoaria* yields the Zedoary; and we have obſerved, that the uſual ſophiſtication of this drug is by mixing Turmeric (CURCUMA *longa*) with it.

Being a native of the Eaſt-Indies, it requires the warmth of a ſtove. Propagated by cuttings.

From the Bankſian Herbarium we learn that it flowered at Spring-Grove in 1793, a year or two before which time, as we are informed by Mr. DRYANDER, it was introduced to this country by Sir GEORGE YONGE, Bart.

Our drawing was taken from a plant, at the Botanic Garden, Brompton.

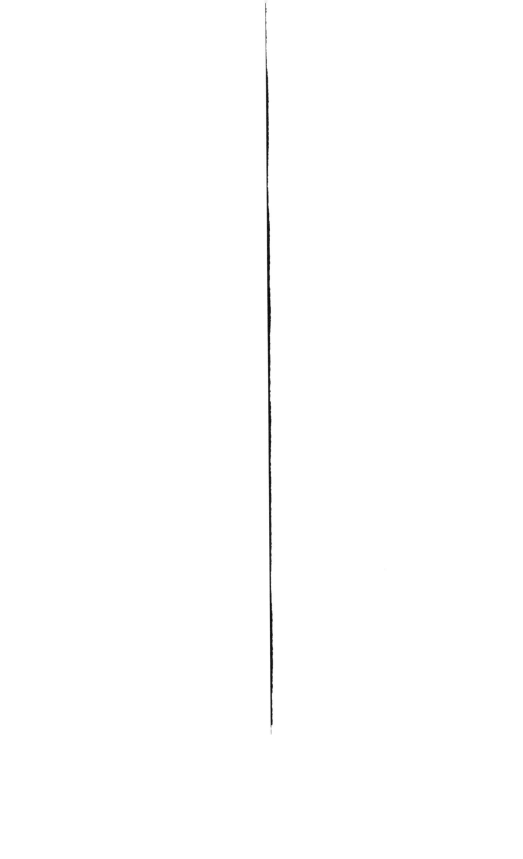

Edwards del. Pub by T Curtis, St Geo Crescent Apr 1 1806 F Sansom sculp

Generic Character.

Cal. inferus. *Cor.* infundibuliformis. *Drupa* 2-fperma.

Specific Character and Synonyms.

STYRAX *lævigatum ;* foliis oblongis utrinque glabris, pedun-
culis axillaribus unifloris folitariis binifve. *Hort.*
Kew. v. 2. *p.* 75. *Willd. Sp. Pl. v.* 2. *p.* 624.
STYRAX *octandrum. L'Herit. Stirp. nov.* 2. *t.* 17.
STYRAX *glabrum. Cavan. Diff.* 6. *p.* 340. *t.* 188. *f.* 1.
Michaux Fl. Bor.-Amer. 2. *p.* 41.
STYRAX *americana. Lamarck Enc.* 1. *p.* 82.
STYRAX *læve. Walt. Car.* 140.

Neither the form of the leaves nor the number of ftamens
will, in this genus, afford permanent diftinguifhing characters ;
but the flowers of this fpecies growing either folitary or in
pairs from the axils of the leaves and at the extremity of the
branches, feems to be conftant. It is a native of the bogs of
Carolina and Georgia in North-America, and fufficiently hardy
to bear the cold of our ordinary winters. Propagated by
layers, and by feeds procured from America. Introduced in
1765.

The ftamens are connected in a ring at the bafe ; on which
account, fome Botanifts have referred this genus to the clafs
Monadelphia.

Our drawing was taken at Mr. LODDIGES, Hackney.

L

ad Edwards del   Pub. by T. Curtis, St Geo Crescent Apr 1 1800   J Sanson sculp

ERYNGIUM ALPINUM. ALPINE ERYNGO.

****

Claſs and Order.

PENTANDRIA DIGYNIA.

Generic Charaɛ̄er.

Flores capitati. *Recept.* paleaceum.

Specific Charaɛ̄er and Synonyms.

ERYNGIUM *alpinum ;* foliis radicalibus cordatis indiviſis, caulinis digitato-laciniatis, capitulis ſubcylindri_cis, involucro pinnatifido frondoſo, paleis tri-fidis. *Hort. Kew.* 1. *p.* 327.

ERYNGIUM *alpinum ;* foliis radicalibus cordatis, caulinis ternatis incifis, involucris ſpinoſo-pinnatis ci-liatis. *Vahl. Symb.* 2. *p.* 46. *Willd. Sp. Pl.* 1. *p.* 1359.

ERYNGIUM *alpinum. Sp. Pl.* 337. *Scop. Carn. n.* 300. *Jacq. Ic. Rar.* 1. *t.* 55. *Villars, Dauph.* 2. *p.* 659. *Allion. Ped. n.* 1284. *J. F. Mill. Icon.* 6.

ERYNGIUM foliis radicalibus petiolatis cordatis, involucro pinnato ciliato. *Hall. Helv. n.* 736.

ERYNGIUM aliud montanum. *Dalech. Hiſt.* 1460. *Ed. Gall.* 2. *p.* 339.

ERYNGIUM cæruleum capitulis dipſaci. *Bauh. Pin.* 386.

ERYNGIUM alpinum làtis foliis, magno capite oblongo cæ-ruleo. *Bauh. Hiſt.* 3. *par.* 1. *p.* 88. *Raii Hiſt.* 386.

ERYNGIUM cæruleum genevenſe. *Lob. Ic.* 2. *p.* 23.

ERYNGIUM planum Matthioli. *Dod. Pempt.* 732. *f.* 2. *Camer. Herb. Ed. Germ.* 229. *Epitome* 449.

In one reſpeɛ̄, the above ſynonymy is remarkable ; it has fallen to the lot of few plants, that have been ſo frequently mentioned by botanical authors, to have undergone ſo little change of name.

JOHN

John Bauhin, in his hiftory, informs us, that he fent this plant to Gesner, whofe figure of it publifhed by Camerarius, though never quoted, is the only one before thofe of Jacquin and F. Miller, except Dalechamp's, which is at all characteriftic of the fpecies. Lobel's figure, fo often copied, was originally but a very indifferent one, and, as mutilated by our Morrison, in his too ufual manner, is no longer applicable to this plant or any other.

This hardy perennial, a native of the Alps in moft of the fouthern parts of Europe, is worthy of a place in every curious garden, where its uncommon form and beautiful colour cannot fail to attraɛt the notice of every beholder. It does not owe its charms to the fplendour of its bloffoms, but to the floral leaves, or involucre furrounding the head of flowers, admired for their feather-like appearance and delicate blue colour; all the upper parts of the plant partake of the fame tint, which becomes in defcending more and more dilute.

Propagated by feeds, or by cuttings of its root. Requires a dry foil, or is apt to perifh from the humidity of our winters. Cultivated by Philip Miller in 1752.

Our drawing was taken at the Botanic Garden, Brompton.

Edward del. Pub by T Curtis, St Geo Crescent Apr 1 1806. F.Sansom sculp.

NEPETA LONGIFLORA. LONG-TUBED CAT-MINT.

Class and Order.

DIDYNAMIA GYMNOSPERMIA.

Generic Character.

Corollæ labium inferius lacinula intermedia crenata: faux margine reflexo. *Stamina* approximata.

Specific Character and Synonyms.

NEPETA *longiflora;* cymis fubquinquefloris, corollæ labio fuperiore bifido: tubo filiformi, foliis cordato-ovatis crenatis rugofis petiolatis.

NEPETA *longiflora;* cymis remotis, pedunculatis unilateralibus, paucifloris; foliis fubfeffilibus, cordato-ovatis, rugofis; corollarum tubo longiffimo. *Vent. Hort. Celf. t.* 66.

NEPETA *Willdenowiana. Adams.*

CATARIA orientalis, folio fubrotundo, flore intenfe cæruleo. *Tourn. Cor.* 13.

DESCR. Suffruticofe. *Stem* divided at the bafe: branches long, weak, generally undivided, fquare with obtufe angles. *Leaves* cordate, oblong-oval, or in young plants fuborbicular, crenate, rugofe, hoary underneath, lower ones on longifh petioles, upper ones fubfeffile. *Bractes* like the leaves, but feffile, and more deeply indented. *Flowers* verticillate, peduncles three to five-flowered: lower ones frequently folitary: upper ones oppofite, but for the moft part turned to one fide. *Calyx* cylindrical, ftriate, hairy, teeth nearly equal, the two fuperior a little longer, minutely ciliated. *Tube* of corolla longer than calyx, filiform, curved: *faux* compreffed, funnel-fhaped, *fides* reflected: *upper lip* divided almoft to the bafe, lobes divaricate: *lower lip* large, crenate, hollow, fpotted with

white

white towards the faux. *Stamens* four : *filaments* twisted :
anthers purple, two-lobed, approximate : *pollen* white. *Style*
longer than corolla : *stigma* bifid, acute.

In young plants the leaves are rounder and less hoary : in
the native specimens indeed the whole plant is more hoary than
when cultivated.

Raised by Mr. LODDIGES from seeds sent from Caucasus
under the name of NEPETA *Willdenowiana,* by which title
specimens were sent to Sir JOSEPH BANKS from the same
country by Count MOUSSIN POUSHKIN; but VENTENAT'S
name having the right of priority, we have adopted it.

The plants in the garden of M. CELS were raised from
seeds gathered on Mount Albours in Persia, by Messrs. BRU-
GUIERE and OLIVIER.

Flowered in the Royal Garden at Kew in 1803.

Propagated by seeds and by cuttings. May be considered
as hardy, but it is safest to give the protection of a frame
during the winter.

Flowers from May through the whole summer.

CURTIS'S
BOTANICAL MAGAZINE;

O R,

Flower-Garden Displayed:

IN WHICH

The moſt Ornamental FOREIGN PLANTS, cultivated in the Open Ground, the Green-Houſe, and the Stove, are accurately repreſented in their natural Colours.

TO WHICH ARE ADDED,

Their Names, Claſs, Order, Generic and Specific Charaĉters, according to the celebrated LINNÆUS; their Places of Growth, and Times of Flowering:

TOGETHER WITH

THE MOST APPROVED METHODS OF CULTURE.

A W O R K

Intended for the Uſe of ſuch LADIES, GENTLEMEN, and GARDENERS, as wiſh to become ſcientifically acquainted with the Plants they cultivate.

CONTINUED BY

JOHN SIMS, M. D.
FELLOW OF THE LINNEAN SOCIETY.

VOL. XXIV.

Invitant croceis halantes floribus horti.
VIRG.

L O N D O N:

Printed by STEPHEN COUCHMAN, Throgmorton-Street.
Publiſhed at No. 3, St. GEORGE'S-CRESCENT, Black-Friars-Road;
And Sold by the principal Bookſellers in Great-Britain and Ireland.
MDCCCVI.

AMARYLLIS ORNATA. (β.) WHITE CAPE-COAST LILY.

✱✱✱✱✱✱✱✱✱✱✱✱✱✱✱✱✱

Claſs and Order.

HEXANDRIA MONOGYNIA.

Generic Charaƈter.

Spatba 1-bivalvis. *Cor.* ſupera tubuloſa infundibuliformis ſexdiviſa, vel hexapetalo-partita ſubrotata ; bilabiatim irregularis, vel regularis ; laminæ ſubæquales ſimiles recurvatæ, vel reƈtiores concavæ incurvulæ. *Stam.* imis laciniis inſita aut per omnem tubum adnata, declinato-aſſurgentia ſubfaſciculata, raro ex ereƈto-divergentibus ſurſum conflexula. *Stylus* tenuis, elongatus. *Stigm.* 1 hians vel 3 recurva. *Capſ.* membranacea, ex oblonga atque trigona ad depreſſo-ſphæricam et pulvinato-toroſam. *Sem.* plura globoſa vel varie preſſa, ſæpius numeroſa plana paleacea, modo ſolitaria carnoſo-baccata, uno ſingulum loculum vel omnem capſulam (cujus reſpondet cavitati) occupante. *G.*

Bulbus tunicatus plexibus ſericeo-filamentoſis membranaceis obvolutus. Folia craſſiuſcula ab anguſto-linearibus ad lato-lanceolata ; ab uno pauciſve et bifariis ad plurima ſparſa ambientia faſciculatim divergentia. Scapus ſubteres, glaber, foliis intermedius vel lateralis. Inflor. 1-multiflora et umbellata. Hic ſæpe accidit quod a germine indice polyſpermi fruƈtus proveniat is tantum cum ſemine ſolitario, quando ex ovulis unum (abortivis reliquis, vel iſto forſan validiore precociuſve fœcundato illiſis) in molem ingrandeſcat cavitati capſulæ vel loculi æqualem. Eſt iſte mos pleriſque plane adventitius, in nonnullis uſitatior, in aliis veriſimiliter conſtans et naturalior. Germinant hæ maſſæ carnoſæ, et ſæpius vireſcentes, modo a ſolito haud alieno ; viſum enim eſt ejus in ornata folium primarium altius ſurgere lobo cum teſta gravatum magnitudine ovi columbini vel majore.

Per præſentem ſpeciem confluit Genus cum Crino diſcrepante duntaxat per corollam ob laminas ad tubum radiantes regulariter hypocrateriformem.

AMARYLLIS orientalis, marginata, ſtriata et Radula, dant alteri diviſioni facile principium, pro qua titulus ad manum habemus ob orientalem jampridem Heiſtero in genus cum nomine BRUNSVIGIÆ *evocatam, licet aliis poſtea fuerit minus reƈte ad* AMARYLLIDEM *redaƈta : qua diſcrepat capſula turbinata, trialatim triloba (molendinacea), ſcarioſo-rigente, ſubdiaphana, ſubſplendente ; ſeminibus paucis et fere aciniformiter promuƈlis ; habitu præterea haud parum, per eum enim plurimum aſſimilatur* MASSONIAM, *a qua rurſus diſtat germine infero, natura ſua bis alvi necnon longius ſcapoſa.* G.*

Specific

AMARYLLIS *ornata (scapo extrafoliaceo)* foliis oblongo-lan‑
ceolatis, undatis, lineatis, minute ciliato-fcabri‑
dis, fafciculatim ambientibus, extimis recum‑
bentibus; floribus feffilibus; corolla fubhypo‑
crateriformi; limbo campanulato-bilabiato;
tubo longiore craffe pedunculoideo in laminas
elliptico-lanceolatas abrupte ampliato; fructu
monofpermo et fubfolido-farcto. *G.*

(α) laminis albis cum difco purpurafcente. *G.*

AMARYLLIS *ornata.* *Hort. Kew.* 1. 418. *Mart. Mill. Dict.*
Willd. Sp. Pl. 2. 55.

AMARYLLIS *zeylanica.* *Sp. Pl.* 421. *L'Herit. Sert. Angl.*
13. *Mart. Mill. Dict. Willd. Sp. Pl.* 2. 56.
Roxburgh. Pl. Coromand. tab. ined. optima.

AMARYLLIS *Brouffoneti.* *Lil. a Redouté, t.* 62.

AMARYLLIS *yuccoides.* *Thompfon's Bot. Difpl. No.* 4. *Pl.* 12.

AMARYLLIS *fpectabilis.* *Bot. Rep. tab.* 390.

AMARYLLIS *bulbifperma.* *Burm. Prod.* 19.

CRINUM *zeylanicum.* *Reich.* 2. 24. *Linn. Syft. Veg.* 263.
id. a *Murr.* 318. *Lour. Flor. Cochin. p.* 198.
(α, β).

LILIO-NARCISSUS africanus, &c. *Ehret. Pict.* 5. *f.* 2.
Trew. Ehret. t. 13.

LILIO-NARCISSUS zeylanicus. *Comm. Hort. Amft.* 1. 73.
t. 73. *Rudb. Elyf.* 2. 191. *f.* 2.

TULIPA Javana. *Rumph. Amboin.* 5. *p.* 306. *c.* 8. *t.* 105.
optime.

(β) laminis albis extus cum aliquo virore fuffufis. *G.*
AMARYLLIS *Jagus.* *Thompfon's Bot. Difpl. No.* 2. *Pl.* 6.

CRINUM *giganteum.* *Bot. Rep. t.* 169.

After a diligent refearch (fully impreffed with a contrary
fufpicion arifing from their far diftant, yet in fact not unanala‑
gous, abodes) we are now fatisfied of the fpecific identity of the
Indian *zeylanica* of LINNÆUS and the African *ornata* of the
Hort. Kew. of which the prefent plant is an obvious variety.
LINNÆUS moft probably took up, as well as named, his fpecies
from the plates and defcriptions of COMMELIN and RUDBECK.
(α) is faid to have been introduced in 1740 by Lord PETRE,
from whofe plant EHRET defigned his plate; but we more than
fufpect the exactnefs of his information, when he ftates it to be
a native of the Cape of Good Hope. Among the Gardeners it

is known by the appellation of " the Cape-Coaft Lily," and was certainly fent to us fome years back by Dr. Afzelius from Sierra Leone; from which colony it was alfo introduced into France by the Botanift whofe name it has received in one of the works above cited. Dr. Roxburgh found fpontaneous fpecimens on the Coromandel-Coaft; Rumpu mentions it as being an inhabitant of the gardens of Amboyna, where it had been received from Batavia, and there known by the name of the Java-Tulip.

Loureiro met with two varieties in China and Cochin-China, and fays, that he found the bulbs anfwer the fame medical purpofes as thofe of the Officinal Squill: from Commelin we learn, that in Holland it was fuppofed to be a native of Ceylon: (β) was received by us alfo, through the means of Dr. Afzelius, from the colony of Sierra Leone; where it is faid to grow in the water (moft probably in fpots that are only periodically inundated) and to be with great difficulty obtained, owing to the jealoufy of the natives, by whom it is held in fuperftitious veneration, being ufed as an amulet or charm to preferve them in war, as well as almoft every other fpecies of danger. Both varieties agree in a decided predilection for low fandy fituations, as well as of water, and we accordingly perceive in our ftoves, that the fize and number of flowers depend much upon the greater or lefs proportion of the latter element that has been fupplied during the time of their vegetation.

In the adoption of the fpecific title of *ornata* in preference to the older one of *zeylanica*, we were influenced firft by its being now as univerfally eftablifhed as the other, and then by its being lefs liable to miflead.

Bulb large (fometimes weighing between three and four pounds) oval-oblong, faftigiate and frequently partly above ground. Leaves from one to three feet in length, feveral, fpringing in a fparfe fafcicle from the crown of the bulb; oblong-lanceolate, channelled-depreffed, fcored, waved, edged with a filiform minutely and fcabroufly ciliate cartilaginous rim, traverfed longitudinally by a broad thick bluntly keeled midrib, narrowed, thickened, and convolute at their bafe, outermoft recumbent, inner narrow, upright. Scape (fometimes two) plano-convex, ftraight, extrafoliaceous, one to three feet or more high. Spathe rather fhorter than tube, bivalved. Umbel two to thirteen-flowered. Flowers large, fragrant, feffile; fubhypocrateriform; limb fubringently campanulate; claws concrete into a tube longer than this, being from four to fix inches or more in length, pedunclelike, thick, fubcylindrically trigonal, ftrict, fucculent, cuniculate, very flightly curved,

curved, fwelling a little towards the germen; mouth naked, abruptly (that is without the ufual gradually enlarged faux) expanding into elliptically-lanceolate broad laminæ, which are recurved; inner rather the wideft.   Stamens filiform, declined-affurgent, fhorter than limb, decurfively adnate to the bore of the tube, from the mouth of which they iffue unattached; anthers linear-oblong, firft yellow, then brownifh, lanceolate, incumbent, balancing.   Germen feffile, oval-oblong, even, fmooth, green, confluent with the tube; Style fomewhat flenderer than the filaments; Stigma fubtrilobately depreffed, hiant, fimbriate.

Thefe plants are now common in our ftoves; flower freely; are eafily propagated and very ornamental.   *G.*

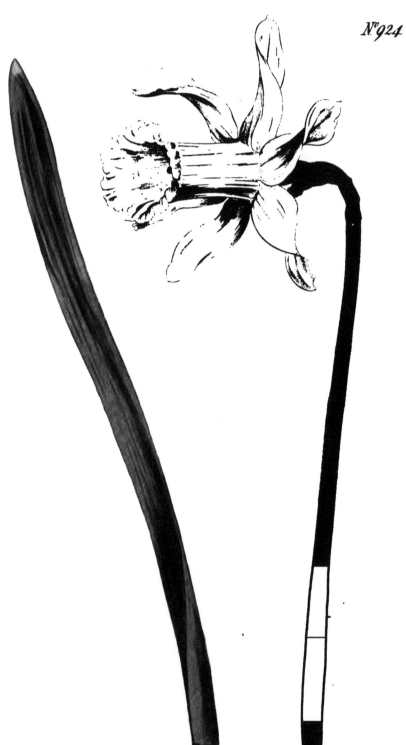

S. A. Edward del. Pub by T. Curtis St Geo Crefcent May 1 1806. F. Santom sculp

NARCISSUS MOSCHATUS (α). WHITE LONG-FLOWERED DAFFODIL.

✳✳✳✳✳✳✳✳✳✳✳✳✳✳✳✳✳

Clafs and Order.

HEXANDRIA MONOGYNIA.

Generic Charaɓer.

Spatba follicularis latere dehifcens. *Cor.* fupera tubo in limbum externum fexpartitum çalycinum et in interiorem fubintegrum corollaceum *(coronam* fi velis) abeunte, hinc infundibuliformis inde hypocrateriformis. *Stam.* tubo varie adnata intra coronam qua breviora. *Stigm.* unum trilobum aut 3 brevia. *Sem.* plura globofa aut varie preffa. *G.—Vid. Obf. Nᵘᵐ.* 925.

Specific Charaɓer and Synonyms.

NARCISSUS *mofcbatus (ftaminibus æqualibus a fundo tubi liberis porreɓo-conniventibus)* foliis loratis concavis cæfiis; flore folitario fubcernuo; tubo turbinato brevi longitudine pedunculi cum germine; laciniis lanceolato-oblongis obliquis; corona reɓo-cylindrica, verticaliter fubplicato-rugofa, iifdem longiore, furfum parum ampliata. *G.*

NARCISSUS *mofchatus. Sp. Pl.* 415. *Mart. Mill. Diɓ. Willd. Sp. Pl.* 2. 37. NARCISSE *mufquè. Lam. Encyc.* 4. 423.

NARCISSUS *cernuus. Rotb. Cat. Bot. fafc.* 1. § 43. *Id. in Ann. Bot.* 2. 25.

PSEUDO-NARCISSUS albo flore. *Cluf. Append. alt. auɓ.* cum *Ic.*

NARCISSUS fylveftris totus albicans minor. *Barrel. Ic.* 945, 946.

NARCISSUS fl. exalbido calyce prælongo fimbriato. *Rudb. Elyf.* 2. 82. *fig.* 18.

PSEUDO-NARCISSUS pyrenæus variformis. *Park. Parad.* 101. *f.* 2.

PSEUDO-NARCISSUS pallidus præcox. *Id. l. c. p.* 99. *abfque Ic.*

PSEUDO-

PSEUDO-NARCISSUS totus albus. *Hort. Eyſt. Vern. Ord. 2.*
fol. 2. fig. 2.

(α) coronæ margine crifpula erofo-dentata. *G.*

(β) PSEUDO-NARCISSUS tubo hexangulari. *Park. Par.*
t. 101. *f* 5.

(γ) PSEUDO-NARCISSUS tubo quafi abfciffo. *Id. t.* 107.
f. 1.

In fpecies evidently taken up from figures, often defective, in which diftinct ones have been confounded together as fynonymous to each other, then trufted to the mercy of a fhort phrafe by way of defcription, it is not a very eafy thing to fay for certain which were the precife plants intended by the author as his ftandards. Such feems to be the cafe in the prefent inftance; for this is certainly one of the plants called in by LINNÆUS to his *mofchatus;* but whether fome other of his fynonyms may not be the plant intended we are at a lofs to fay. Is this fpecies really diftinct from *bicolor?* Are the fynonyms added as varieties really plants of the fame fpecies? we have not yet met with more than this one of them in any of our collections. *Mofchatus* has efcaped the *Hortus Kewenfis,* nor have we ever feen a figure of it in any recent work. Differs from *bicolor* in having a crown more truly cylindric, lefs cleft and not fo widely or far patent; from both that and *Pfeudo-Narciffus,* by its drooping flower, and from the latter by its longer crown and ftamens reaching only to about the middle of that; from both again in colour and fcent. The pedicle is curved and enveloped by the fpathe. We fuppofe it has received its name from fome of the varieties having been defcribed by the old Botanifts as poffeffing flowers " *cum mofcari odore.*"

This has little fcent of any fort; but what it has is pleafant, fomewhat like ginger, and not in the leaft refembling that of mufk. Blooms early in April. Hardy.

Our drawing was taken at the Botanic-Garden, Brompton. *G.*

ERRATA.

Edwards del. Pub. by T. Curtis, S.ᵗ Geo: Crescent May.1.1806. F. Sanṡom sculp

NARCISSUS TAZETTA. POLYANTHUS NARCISSUS.

✷✷✷✷✷✷✷✷✷✷✷✷✷✷✷✷✷

Claſs and Order.

HEXANDRIA MONOGYNIA.

Generic Charaƈter.—Vid. ſupra N. 924.

OBS. *Bulbus tunicatus tegminibus membranaceis. Folia plura, bifaria, lineari-lorata, ſubſucculenta, craſſiuſcula, plana, ſupra parum depreſſa infra carinato-coſtata, per exceptionem canaliculato-ſemiteretia, juncea. Scapus nudus centralis, ſæpius compreſſus vel anceps. Flores flaveſcentes aut albicantes vel ex utroque more bicolores. Stylus triqueter, quaſi tres conglutinati. Capſ. membranacea. Differt* PANCRATIO *ſtaminibus intra coronam (hic haud ſtaminilegam) dilapſis.* G.

Specific Charaƈter and Synonyms.

NARCISSUS *Tazetta (ſtaminibus in æqualibus, 3 incluſis adnatis, 3 ſubadnatis tubo emicantibus)* foliis depreſſo-planis, loratis, obtuſe attenuatis ; ſpatha multi-flora ; corona cupulata integriori laciniis ovato-lanceolatis explanatis triplo breviore ; tubo trigono his ſubæquali pedicellis ereƈtis breviore. G.

NARCISSUS *Tazetta. Sp. Pl.* 416. *Syſt. Veg. Murr.* 317. *Hort. Kew. v.* 1. *p.* 410. *Willd. Sp. Pl.* 2. *p.* 39. *Quer. Flor. Eſpan. v.* 5. *p.* 477. *Haworth in Linn. Tranſ. v.* 5. 245. *Desfont. Flor. Atl.* 1. 282. *Brotero Flor. Luſit.* 1. 551. *Lil. a Redouté t.* 17.

NARCISSUS africanus aureus majór. *Park. Par.* 81. *f.* 1. *Floril. Auƈt.*

NARCISSUS africanus luteus minor. *Id. eod.* 81. *f.* 2.

NARCISSUS luteus polyanthos africanus. *Bauh. Pin.* 50.

NARCISSUS latifolius flore prorſus flavo. *Cluſ. Hiſt.* 156.

Native of Spain, Portugal and moſt probably of the coaſt of Barbary ; now one of the commoneſt ornaments of our gardens, having

having been cultivated here from the time of PARKINSON and GERARD. The beſt roots are annually imported by the Seedſmen from Holland, where two or three hundred varieties are enumerated ; but we ſhall defer to a future number our obſervations on the ſpecies, and on ſuch plants as we conſider really varieties of it or not. Thoſe that wiſh to be informed minutely of the mode of cultivating and raiſing it from feed, have only to refer to MILLER's Dictionary.

Tazzetta means a ſmall cup, and is the name given to theſe flowers in Italy from the ſhape of their crown. LINNÆUS has ſpelt it with one z inſtead of two.

The roots of this ſpecies are ſometimes the ſize of both the fiſts put together. The ſtem and leaves ſometimes two feet or more high. The ſcent is pleaſant, but very ſtrong, even pungent. Blooms in the open ground about April. Forces well in water, ſand, or common loam. We have not added the ſynonym from THUNBERG's *Flora Japonica*, as we cannot always rely on his accuracy in this department. His may be the ſame plant as ours. G.

Syd.Edwards del Pub by T.Curtis. S.t Geo Crefcent May 11806 F Sanfom sculp

POLYANDRIA TRIGYNIA.

Generic Character.

Cal. 5-phyllus. *Petala* 5, regularia. *Germina* 2—5. *Styli* O. *Capf.* polyfpermæ.

Specific Character and Synonyms.

PÆONIA *tenuifolia ;* foliis biternatis, foliolis multipartitis nudis, laciniis lineari fubulatis, capfulis tomentofis. *Willd. Sp. Pl.* 2. *p.* 1223.

PÆONIA *tenuifolia ;* foliolis linearibus multipartitis. *Linn. Sp. Pl.* 748. *Syft.* 502. *Reich.* 2. 610. *Del.* 9. *t.* 5. *Gmel. Sib.* 4. *p.* 185. *t.* 73. *Pall. Roff.* 2. *p.* 95. *t.* 87. *Zinn. Goett.* 127. *Gært. Fruct.* 1. 309. *t.* 65. *f.* 1. *Mart. Mill. Dict. n.* 5. *Hort. Kew.* 2. *p.* 241. *Meerb. ic.* 25.

As moft Botanifts agree that the more ufual number of germens in this genus is three, we have ventured to remove it from the fecond order, where it has hitherto been arranged, to the third, and this we have done for the fake of affociating it with DELPHINIUM and ACONITUM, to which it is nearly allied by nature.

This very ornamental flower is perfectly hardy, thriving almoft in any foil or fituation. Grows naturally in the Ukraine and about the precipices on the borders of the Tanais, the Volga, and the Terek. The flowers of the wild plant are far lefs fpecious than when cultivated.

Our drawing was taken at the Botanic-Garden, Brompton. Flowers in May and June. Is eafily propagated by parting its roots or by feeds.

❶

Syd. Edwards del Pub. by T. Curtis, St Geo Crescent May 1.1806 F Sansom sculp

Cor. campanulata, fundo clauſo valvis ſtaminiferis. *Stigma* 3-fidum. *Capſ.* infera, poris lateralibus dehiſcens.

Specific Character and Synonyms.

CAMPANULA *collina;* foliis ovato-lanceolatis crenulatis in petiolum decurrentibus; panicula laxa ſub-fecunda, corollis cyathiformibus; laciniis revolutis piloſis, foliolis calycinis erectis glabris corollis multo brevioribus.

DESCR. *Root* perennial? *Stalk* erect, angular, hairy at the lower part: *branches* few, erect, generally three-flowered. *Flowers* looking one way, nodding, peduncled, ſhewy, bright blue. *Calycine leaflets* ſimple, ſmooth, diſtant, not reflexed: the part adhering to the germen three-grooved, angles clothed with a few white hairs looking downwards. *Tube* of corolla cup-ſhaped nearly hemiſpherical: *laciniæ* pointed, rolled back, hairy round the margin on the innerſide. *Valves* ovate, acute, conniving: *filaments* very ſhort: *Anthers* long, linear, ſpeedily waſting. *Germen* top-ſhaped: *Style* erect, ſomewhat longer than tube: *Stigma* trifid, revolute. *Capſule* three-celled.

Seeds of this new ſpecies of Campanula were received from Caucaſus by Mr. LODDIGES, of Hackney, under the name which we have adopted; we find ſpecimens of the ſame plant and under the ſame name, but mixed with a different ſpecies, in the collection ſent from that country to Sir JOSEPH BANKS, from Count MOUSSIN POUSHKIN. It flowers with us in July and Auguſt. Is probably hardy enough to bear the cold of our winters, but will be more certainly preſerved, by being kept under a frame with other alpine plants.

Syd Edwards del   Pub by T Curtis, S.t Geo Crescent May 1 1806   F Sansom sculp

VACCINIUM BUXIFOLIUM. BOX-LEAVED WHORTLE-BERRY.

****

Clafs and Order.

OCTANDRIA MONOGYNIA.

Generic CharaĉZer.

Cor. 1-petala. *Stamina* receptaculo inferta. *Antheræ* apice poris 2. *Bacca* infera, 4-locularis, polyfperma.

Specific CharaĉZer and Synonyms.

VACCINIUM *buxifolium ;* floribus decandris fafciculatis racemofifve axillaribus, pedunculis braĉteatis, ftigmatibus capitatis, foliis ovalibus crenulatis planis impunĉtatis.

VACCINIUM *buxifolium ;* foliis obovatis, dentatis, glabris, fubtus æquatis : fpicis e fuperioribus axillis, denfe multifloris : ftigmatibus hemifphæricis. *Salifb. in Parad. Lond.* 4.

VACCINIUM *brachycerum ;* pumilum : foliis *Buxi,* obovalibus, rariter manifefteque crenatis : fafciculis fubfeffilifloris : corolla brevi : filamentis glandulofis ; antheris breviffime corniculatis. *Michaux Flor. Bor. Am. v.* 1. *p.* 234.

To this beautiful dwarf fpecies of Vaccinium we apply the name given it in the Paradifus Londinenfis, in preference to that of MICHAUX, though the latter has the right of priority ; not merely becaufe when accompanied with a good figure a name is likely to be more generally adopted ; but alfo becaufe we are not without fome doubts of the identity of our plant and that of MICHAUX. In habit it approaches to VACCINIUM *Vitis Idæa,* but the corolla is urceolate, with a minute fivecleft border, the anthers included, and have their opening not at the extremity but on one fide ; ftigma capitate ; leaves crenulated, not dotted underneath, nor with the edges rolled back.

Our drawing of this very rare fhrub was taken at Mr. WOODFORD's, late of Vauxhall, who purchafed it from the colleĉtion of the late Mr. SYKES, of Hackney. A native of North-America. Flowers in April or early in May. Requires the fame treatment as the hardy heaths, and may be propagated by layers.

Syd Edwards del   Pub by T Curtis St Geo Crescent May 1 1806 F Sansom sculp

Corollæ limbus tubulato-ventricofus : fauce claufa radiis fubulatis.

Specific Character and Synonyms.

SYMPHYTUM *afperrimum ;* caulibus aculeatis foliis ovalibus acutis pedunculatis : floralibus oppofitis, racemis geminis.
SYMPHYTUM *afperrimum.* *Donn. Hort. Cantab.*
SYMPHYTUM *orientale,* folio fubrotundo afpero, flore cæruleo. *Tournef. Cor.* 7.

This fpecies of Symphytum, a native of Caucafus, is by far the largeft of the genus, growing to the height of five feet, and is really an ornamental, hardy perennial, which will thrive in any foil or fituation. It differs from SYMPHYTUM *orientale* not only in ftature and in the greater roughnefs of the leaves, but in the ftems being not merely hifpid, but covered with fmall curved prickles; the floral leaves are conftantly oppofite, which is feldom the cafe in *orientale*. The nectaries in both are flat, not fiftulous.

According to Mr. DONN, it was introduced in 1801, we believe, by Mr. LODDIGES, of HACKNEY.

Our drawing was taken at the Botanic Garden at Brompton, where we have obferved it fome years in the greateft vigour.

Propagated by parting its roots or by feeds.

Syd! Edwards del.   Pub! by T. Curtis, N! 6rc. Walworth June 1 1806   F. Sansom sculp

LIMODORUM ALTUM. TALL LIMODORUM.

****

Class and Order.

GYNANDRIA DIANDRIA.

Generic Character.

Nectarium monophyllum, concavum, pedicellatum intra petalum infimum.

Specific Character and Synonyms.

LIMODORUM *altum;* floribus imberbibus, fpicis fubpaniculatis. *Hort. Kew.* 3. *p.* 301. *Mart. Mill. Dict. n.* 2.

LIMODORUM *altum. L'Herit. Sert. Ang.* 28.

HELLEBORINE americana, radice tuberofa, foliis longis anguftis, caule nudo, floribus ex rubro pallide purpurafcentibus. *Mart. Cent.* 50. *t.* 50. *Mill. Ic. t.* 145.

LIMODORUM *altum. Syft. Veg. Murr.* 816 ? *Sp. Pl. Reich.* 4. *p.* 32 ? *Swartz. Obf.* 323 ?

HELLEBORINE radice arundinacea, foliis ampliffimis lyratis. *Plum. Ic.* 189 ?

SATYRIUM 10, foliis liratis longiffimis, fcapo florifero partiali, fubfquamofo. *Brown Jam. p.* 325 ?

It is not altogether certain that the plant, here figured, is the real LIMODORUM *altum* of LINNÆUS. The fpecimen in the Bankfian Herbarium, marked as correfponding with that in the Linnean, has a fimple, erect fcape, anfwering very well to PLUMIER's figure above referred to: whereas the flowering ftem in this is, for the moft part, branched a confiderable part of its length. As it is however undoubtedly the fpecies meant to be characterized in the Hortus Kewenfis and in MARTYN's MILLER's Dictionary, in a cafe of uncertainty, we think it fafeft to retain the name by which it has long been known in our gardens; expreffing our doubt of the fynonyms, which,

agreeing

agreeing better with the specimen in the Linnean Herbarium, may perhaps belong to another species. This doubt attaches in some degree to the description by SWARTZ, according to which the scape is simple, or only a little divided at the upper end. This author also describes the nectarium, as being furrowed or grooved, whereas, in our plant, it is marked with about seven yellow, raised ridges. The leaves vary so much from linear-lanceolate, to broad-lanceolate, that nothing certain can be determined by them. In other respects SWARTZ's description accords with this species. The fruit might perhaps decide the question, but unfortunately with us the flowers drop off without producing any : from the above-mentioned specimen it appears, that, as soon as the flower fades, the peduncle is reflected and the fruit becomes depending, which is well represented in PLUMIER's drawing, though nearly omitted in the published engraving.

MILLER evidently confounds the *tuberosum* with this species, when he says that he had received roots of it from Philadelphia and the Bahama Islands ; and L'HERITIER certainly misapplied the synonyms of this author and of MARTYN, above quoted, in which error he has been followed by SWARTZ ; but the bearded nectarium of *tuberosum* will always distinguish it : indeed the very different form of its resupinate corolla and the greater length of the anther-bearing column (see No. 116 of this work) may even lead to a doubt if it really belong to the same genus as *altum*.

The LIMODORUM *altum* sometimes varies with white, and pale rose-coloured flowers ; as the latter variety generally grows to a larger size, it has by some been suspected to be a distinct species, but for this we see no good grounds.

A native of Jamaica, but found only in the cooler parts of the mountains, in dry stony and sandy situations. It is not therefore very impatient of moderate cold, but frequently suffers from the moist heated atmosphere of the bark stove. Flowers in May, June, and July. Is easily propagated by offsets from tuberous roots.

The plant from which our drawing was taken was received from Mr. LODDIGES of Hackney.

S.J. Edwards del. Pub. by T.Curtis, S.t Geo: Crescent June 1 1806. F. Sansom sculp

PHYTOLACCA DECANDRA. VIRGINIAN POKE.

****

Class and Order.

DECANDRIA DECAGYNIA.

Generic Character.

Cal. o. *Petala* 5 calycina. *Bacca* fupera 10-locularis, 10-fperma.

Specific Character and Synonyms.

PHYTOLACCA *decandra;* floribus decandris decagynis. *Sp. Pl.* 631. *Willd.* 2. *p.* 822. *Reich.* 2. *p.* 406. *Blackw. Ed. Germ. t.* 515. *Mill. Ill. Zorn. Ic. Mart. Mill. Dict. n.* 3. *Gron. Virg.* 161. *Desf. Atl.* 369. *Hall. Helv. n.* 1007. *Abbot. Georg. t.* 97. *Michaux Fl. Bor-Am.* 1. *p.* 278.

PHYTOLACCA vulgaris. *Dill. Elth.* 318. *t.* 239. *f.* 309.

SOLANUM magnum virginianum rubrum. *Park. Theat.* 347. 8. *f.* 3. *Morif. Hift.* 3. *p.* 522. *f.* 13. *t.* 1. *f.* 1.

SOLANUM racemofum americanum. *Raii Hift.* 662. *Pluk. Phyt. t.* 225. *f.* 3.

SOLANUM racemofum tinctorium americanum, foliis et feminibus Amaranthi. *Herm. Hort. Lug.* 574. *Weinm. Phyt. t.* 936.

BLITUM americanum. *Munting. Icon.* 112.

In large gardens, where the room it neceffarily takes up can be fpared, this, in feveral refpects, fingular plant may be allowed a place; for, bearing flowers and fruit at the fame time, it is by no means void of beauty. Being of the natural order of ATRIPLICES of JUSSIEU, the HOLORACEÆ of LINNÆUS, one is not furprifed to find that it is fometimes eaten boiled as fpinach, at the fame time it may be obferved that it has rather a fufpicious afpect, and we are told that in America the root is in common ufe as a domeftic purge, and that two fpoonfuls of white wine, in which an ounce of the dried root has
been

been infufed, will operate as a mild emetic, the more commendable, as the wine is faid to be very little changed in tafte by it. Another fpecies of the fame genus, called Spanifh *Calaloe*, and cultivated in kitchen gardens in Jamaica, as a palatable, wholefome green, is faid by THUNBERG to be in Japan extremely poifonous, though, according to KÆMPFER, cultivated there for the fake of its very nutritious root.

It is remarkable for the different countries in which it is indigenous; Spain, Portugal, Switzerland, Barbary, Virginia, New-England, and Jamaica; perhaps, however, it has been originally imported to Europe from America.

The berries afford a beautiful colour, if it could be rendered durable. They are faid to have been at one time much ufed in Portugal, to give a deep colour to the Red Port; but the tafte being complained of by the merchants, the government ordered the plant to be every where cut down before the berries were ripened.

The number of ftamens, of which we find twelve more generally than ten, will hardly fuffice to diftinguifh this fpecies: the number of ftyles is ftill more indefinite.

A hardy perennial, but faid to be fometimes deftroyed by fevere froft. Propagated by parting its roots or by feeds. Flowers in July and through the latter part of the Summer and Autumn. Cultivated by PARKINSON in 1640, by RAY, in his garden at Cambridge, and by MORISON, the latter of whom has given a better figure of it than he frequently does. Our drawing was taken at Mr. SALISBURY's Botanic Garden, at Brompton.

Syd Edwards del   Pub. by I.Curtis, St Geo Crescent June 1 1806   F. Sansom sculp

SYNGENESIA POLYGAMIA SEGREGATA.

Generic Character.

Cal. 76. 1-florus. *Coroll.* tubuloſæ, hermaphroditæ. *Recept.* ſetoſum. *Pappus* obſoletus.

Specific Character and Synonyms.

ECHINOPS *Ritro;* capitulo globoſo, foliis pinnatifidis ſupra
glabris. *Syſt. Veg.* 797. *Reich.* 3. *p.* 946. *Hort.
Kew.* 3. *p.* 281. *Mart. Mill. Dict. Icon. t.* 130.
ECHINOPS *Ritro. Sp. Pl.* 1314. *Hort. Upſ.* 248. *Villars
Dauph.* 3. *p.* 265. *Scop. Carn.* (ECHINOPUS)
994.
ECHINOPS foliis ſupra glaberrimis, ſubtus tomentoſis caule
multifloro corymboſo. *Gouan Illuſtr.* 74.
ECHINOPS caule ſubunifloro, foliis duplicato-pinnatifidis,
foliolis latiuſculis vicinis. *Gmel. Sib.* 2. *p.* 100.
ECHINOPUS minor. *Baub. Hiſt.* 3. *p.* 72. *Tourn. Inſt.* 463.
CARDUUS ſphærocephalus cæruleus minor. *Baub. Pin.* 381.
Park. Parad. p. 332. *t.* 331. *f.* 5. *Raii Hiſt.* 383.
RITRO floribus cæruleis. *Lob. Icon.* 2. *p.* 8.
CROCODYLIUM monſpelienſium. *Dalech. Hiſt.* 1476.

Two ſpecies of this genus are frequently to be met with
cultivated in our gardens, viz. *ſphærocephalus* and *Ritro.* Both
remarkable for the exact globular form of the flowering
heads.

Our plant, though leaſt common, is the moſt ornamental,
on account of its blue flowers, and better fitted for the flower-
garden from its more moderate ſize.

It

It deferves a place in every Botanift's garden on another account, as affording an obvious example of the order *polygamia fegregata* in the clafs *fyngenefia*, of which order very few examples occur.

A hardy perennial, eafily propagated by parting its roots, which creep under ground, or by feeds.

A native of Siberia and Southern Europe. Cultivated by PARKINSON in 1629, who gives rude figures both of this and *fphæro-cephalus* in his Garden of Pleafant Flowers.

Syd Edwards del. Pub by T Curtis St Geo Crescent June 1 1806. F Sansom sculp

Cor. 4-fida five 4-petala. *Antheræ* lineares, infertæ petalis infra apicem. *Cal.* proprius o. *Nux.* 1-fperma, fupera.

Specific Character and Synonyms.

PROTEA *mucronifolia;* foliis lineari-lanceolatis mucronatis epunctatis planiufculis, bracteis involucri angufte cuneatis integerrimis. *Salifb. in Parad. Lond.* No. 24.

In the delicate velvety white involucrum, with which the head of flowers is furrounded, together with the ftrong con- traft of the red anthers upon the feather-tipped fnowy petals, confifts the chief beauty of this fpecies: of which we do not find any account previous to that of Mr. SALISBURY in the Paradifus Londinenfis.

This author has very properly remarked the near affinity that exifts between this and PROTEA *rofacea,* his *acuifolia.*

Our drawing was made in October laft from a fine plant in Mr. HIBBERT's collection at Clapham. It is a native of the Cape of Good Hope; requires the protection of a greenhoufe, and a treatment fimilar to the reft of the genus.

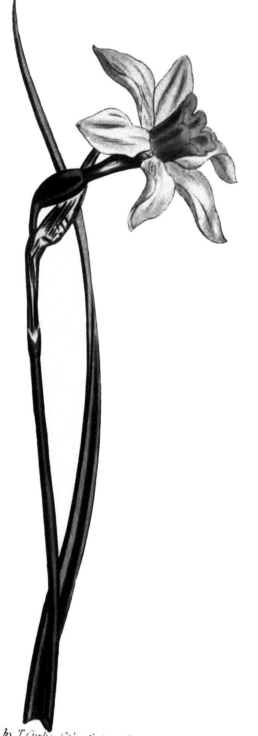

G.J Edwards del.   Pub by T Curtis, Nᵒ Geo Crescent June 1 1806.   F Sansom sculp

NARCISSUS CALATHINUS (α). GREAT YELLOW JONQUIL.

✴✴✴✴✴✴✴✴✴✴✴✴✴✴✴✴✴✴

Class and Order.

HEXANDRIA MONOGYNIA.

Generic Character.—Vid. N°s. 924 & 925.

Specific Character and Synonyms.

NARCISSUS *calathinus (stamina æqualia, tubi medio tenus adnata)* 1—4 flora; scapo tereti, lævi; foliis canaliculato-semiteretibus, sensim attenuatis; crassis; laciniis tubo turbinato-tereti longioribus; corona cyathiformi, læviori, sexlobo-fissa, integra, his sub una tertia parte breviore; stylo hanc subæquante. *G.*

NARCISSUS *calathinus. Sp. Pl.* 415. *Syst. Veg.* 336. *Willd. Sp. Pl.* 2. 39. *Brotero Flor. Lusit.* 1. 551. ?

NARCISSUS *odorus. Sp. Pl.* 416. *Reich.* 2. 19. *Hort. Kew.* 1. *p.* 410. *Haw. in Linn. Transf.* 5. 244. *Curt. Mag. supra* 78. *Willd. Sp. Pl.* 2. 38. rejectis passim synonymis *Am. Acad.* 4. 311 et *Gouan Illustr.* 23. ad NARCISSUM *incomparabilem* spectantibus.

NARCISSUS *Jonquilla* (major). *Quer Flor. Espan.* 5. 477.

NARCISSUS juncifolius max. amplo calyce. *Park. Par.* 89. *f.* 5.——luteus magno calyce. *Id. l. c.* 93. *f.* 4.

NARCISSUS angustifolius flavus magno calyce. *Bauh. Pin.* 51. *Rudb. Elys.* 2. 60. *f.* 5.

NARCISSUS IX. f. angustifolius I. *Clus. Hist.* 1. 158.

NARCISSUS juncifolius præcox major. *Hort. Eyst. Vern. Ord.* 3. *fol.* 7. *f.* 2.

(γ) NARCISSUS polyanthos flore minore stellato toto luteo. *Rudb. l. c.* 6. *f.* 5. ? a LINN. citatus.

Whoever will be at the trouble of turning to the description and synonymy of NARCISSUS *odorus* first taken up in
Amænitates

Amænitates Academicæ, will foon perceive that the prefent fpecies was not the one there intended, but that it was the NAR-
CISSUS *incomparabilis* of this work, which is alfo the *odorus* of
GOUAN. LINNÆUS, in the fecond edition of his *Species
Plantarum,* while he cites the *odorus* of *Amæn. Acad.* evidently lofes fight of that plant, and changes his fpecific phrafe
and defcription, as well as fynonymy, to fuit the fpecies we
have now before us; not aware that it is diftinct from the
one he is incorporating it with, nor that he has already taken
the fame up under the name of *calathinus* in this very work,
moft probably from figures only, as he defcribes the leaves
flat. REICHARD afterwards added the fynonyms of GOUAN
and HALLER; the latter had himfelf cited *calathinus* to his
plant, with a mark of doubt to the words " foliis planis."

Since one of the three fpecific names fhould now merge, we
have fuppreffed that of *odorus, incomparabilis* being better eftablifhed for the one fpecies, and *calathinus* having been applied
to this only, and being befides more conformable to LINNÆUS's
rules for felecting trivial names.

In weak, young, or even many-flowered plants, the crown is
proportionately fhorter, and the lobes often gnawn or crenulate,
both which characters are loft when the fame plants grow ftronger
or blow with a fingle flower. Called *calathinus* by LINNÆUS,
from the crown refembling a chalice.

A native of the South of Europe. Hardy. Sweet-fcented;
but not fo much fo as others of the genus. Varies with very
double flowers, and is then called by fome Gardeners " Queen
Anne's Jonquil." Blooms in April, have never feen it with
more than four flowers, and but rarely with fo many. *G.*

CORRIGENDA & ADDENDA.

No. 78. For " NARCISSUS ODORUS" read " NARCISSUS CALATHINUS (β)"
and refer to this Number.

No. 121. NARCISSUS INCOMPARABILIS.—Add the following Synonyms,
NARCISSUS *odorus. Am. Acad.* 4. 311. *Gouan Ill.* 23.
NARCISSUS *Gouani. Roth in Ann. of Bot.* 2. 26.
NARCISSUS *Pfeudo-Narciffus.* γ. *Mart. Mill. Dict.*
NARCISSUS *albic. cal. aureo,* &c. *Barrel. Ic.* 927, 928.
NARCISSUS *incomparabilis fl. pl. partim flavo partim croceo. R. Par.*
NARCISSUS *montanus albus apophyfibus præditus. Park. Par.* 71. *f.* 5. NAR-
CISSUS *mattenfe. Id. l. c.* 71. *f.* 2. N. *montanus five nonpareille
totus albus. Id. l. c. f.* 6.

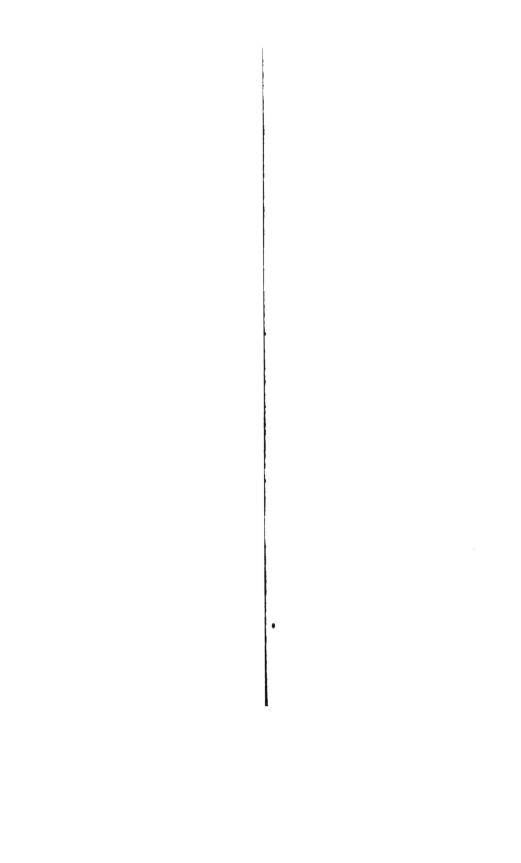

Syd. Edwards del. Pub. by T. Curtis, Glazed Crescent. June 1 1806. F. Sansom.

ORNITHOGALUM UNIFOLIUM. ONE-LEAVED STAR OF BETHLEHEM.

****

Clafs and Order.

HEXANDRIA MONOGYNIA.

*Generic Charaĕler.—Vid. N*ᵘᵐ· 918.

Specific Charaĕler and Synonyms.

ORNITHOGALUM *unifolium* folio folitario fcapum excedente, altius vaginante, cufpide longa
compreffo-tereti ftriata caudatim termi-
nato; floribus paucis fpicatim feffilibus;
braĕleis membranaceis lato-naviculari-
bus; filamentis æqualibus planis fubulato-linearibus; germine fubftipitato
obconico-trigono, angulis fulcatis. *G.*
ORNITHOGALUM *nanum. Brotero Flor. Lufit.* 1. 529.
SCILLA *unifolia. Sp. Pl.* 443. *Link et Hoffm. de Scilla in
Ann. Bot.* 1. 106. *Willd. Sp. Pl.* 2. 131.
ORNITHOGALUM fpicatum unifolium et trifolium flore
niveo odorato. *Grifl. Lufit. V. L. n.* 1596.
BULBUS μονοφυλλος. *Cluf. App. Alt. cum Ic.*
BULBUS monophyllus. *Baub. Hift.* 2. 622.

This fingular little vegetable grows in the greateft abundance
on the barren hills and wilds of Portugal, whence it was received by CLUSIUS, who defcribed and caufed it to be engraved in his great work; but fince his time we do not know
of any book in which there is an original figure of it; nor is
it enumerated among the plants of the *Hortus Kewenfis.*

Our drawing was taken from a fpecimen that flowered in
March laft with Mr. RUDGE author of PLANTÆ GUIANENSES,
a work containing delineations of many rare or till now unknown vegetables, the plates of which are engraved from defigns
executed by his lady with great fkill and accuracy.

Bulb

Bulb about the fize of a nutmeg, ovate. Leaf folitary, From eight inches to a foot in length, narrow-lorate, concave, ftriated without, far-fheathing, caudately terminated by a long compreffedly round ftrict cufpis, recurved. Scape fhorter than leaf, roundifh, thickened upwards; flowers white, three to five, feffile, fpiked; bractes membranous, broad-cymbiform, pointed, reaching half-way up the flower or further, keel green; corolla patent, fegments oblong, fomewhat tranfparently ftreaked, inner ones rather the broadeft and lefs expanded; organs about one-third fhorter than the fegments which are nearly equal; filaments flat, fubulate-linear, equal; ftyle fetaceous, longer than germen; ftigma fimple, pubefcent; germen fubftipitate, obovate-oblong, three-lobedly angular, angles furrowed. By CLUSIUS and others it is defcribed as fweet-fcented, by BROTERO as fcentlefs, and fo it feemed to us. Should be kept in a garden-frame during winter.

The root was brought from Portugal by Sir THOMAS GAGE, who has kindly propofed to affift us in procuring the rarer plants of this order and the Enfatæ indigenous of that country, all which are now nearly loft to our gardens. *G.*

Bulb about the fize of a nutmeg, ovate. Leaf folitary, from eight inches to a foot in length, narrow-lorate, concave, ftriated without, far-fheathing, caudately terminated by a long compreffedly round ftrict cufpis, recurved. Scape fhorter than leaf, roundifh, thickened upwards; flowers white, three to five, feffile, fpiked; bractes membranous, broad-cymbiform, pointed, reaching half-way up the flower or further, keel green; corolla patent, fegments oblong, fomewhat tranfparently ftreaked, inner ones rather the broadeft and lefs expanded; organs about one-third fhorter than the fegments which are nearly equal; filaments flat, fubulate-linear, equal; ftyle fetaceous, longer than germen; ftigma fimple, pubefcent; germen fubftipitate, obovate-oblong, three-lobedly angular, angles furrowed. By Clusius and others it is defcribed as fweet-fcented, by Brotero as fcentlefs, and fo it feemed to us. Should be kept in a garden-frame during winter.

The root was brought from Portugal by Sir Thomas Gage, who has kindly propofed to affift us in procuring the rarer plants of this order and the Enfatæ indigenous of that country, all which are now nearly loft to our gardens. *G.*

Pub by T Curtis, St Geo Crescent June 1 1806

LILIUM SUPERBUM. SUPERB LILY.

✸✸✸✸✸✸✸✸✸✸✸✸✸✸✸✸✸

Claſs and Order.

HEXANDRIA MONOGYNIA.

Generic Character.

Cor. infera, hexapetalo-partita; laciniæ deorſum. turbinatim convergentes, intus ſulco longitudinali nudo aut ciliato exaratæ; laminæ varie lanceolatæ, campanulato-digeſtæ, de erectis uſque revoluto-reflexas. *Stylus* in *Stigma* capitato-trigonum clavato-finiens. *Capſ.* oblonga aut turbinato-trigona, cartilaginea. *Sem.* numeroſa, plana. G.

Radix bulbus ſquamoſus. Caulis folioſus, ſimplex. Folia modo tam caulina quam radicalia modo caulina ſola, hæc ſparſa confertiora aut verticillata remotiora. Inflor. terminalis de uniflora uſque corymboſe thyrſoidee vel umbellatim racemoſo-multifloram; pedunculi longiores, nunc ramiformes; bracteæ foliiformes; flores majuſculi ſpecioſi de erectis uſque cernuos. Sem. in quoque loculo ordine gemino per ſtrata in columnas congeſta. Differt FRITILLARIA, *cui proximum vicinum, bulbo verius ſquamoſo; ſummis foliis nequaquam ultra flores comoſo- vel ſubcomoſo-protenſis, corollæ laciniis haud baſi extus toroſis, tum nectarii figura diverſa.* G.

Specific Character and Synonyms.

LILIUM *ſuperbum* bulbo candidiſſimo; foliis omnibus caulinis, lineari-lanceolatis, trinerviis, nudis, glabris, inferioribus verticillatis atque internodiis duplo-longioribus, ſuperioribus ſubſparſis; floribus umbellatim aut thyrſoideo-racemoſis, pendulo-cernuis, laminis revoluto-reflexis. G.

LILIUM *ſuperbum. Sp. Pl.* 434. *Hort. Kew.* 1. 430. *Lam. Encyc.* 3. 536. *n.* 8. *Thornton's Illuſtr. No.* 2. *Willd. Sp. Pl.* 2. 88. *Redoute Lil. t.* 103. excluſo paſſim ſynonymo *Mill. Dict. n.* 8 ad plantam europæam ſpectante.

LILIUM *carolinianum. Michaux Flor. Bor-Amer.* 1. 197.

LILIUM foliis ſparſis multiflorum &c. &c. *Trew. Ehret.* 2. *t.* 11.

LILIUM five Martagon canadenſe &c. *Cateſb. carol.* 2. 56. *t.* 56.

MARTAGON canadenſe majus. *Trew. Seligm. v.* 1. *t.* 26.

This

This splendid native of North-America was introduced by Mr. PETER COLLINSON, from Pennsylvania, about the year 1738. . MICHAUX found it growing in moist grassy spots in Carolina. Spontaneous specimens have seldom more than three flowers in a kind of umbel; but cultivated carefully, and kept in a moist shady border of bog-earth, it will rise to the height of five feet and produce a thyrse of from twelve to fifteen flowers. Differs from L. *Martagon* in having a bulb as white as ivory, not of a reddish-yellow; in having narrower, linear-lanceolate, tender, not obovate-lanceolate subcorrugately veined harsh leaves; has also much shorter internodes. The plant adduced by LINNÆUS and all his successors from MILLER's work, by way of a synonym to this, is quite a distinct species, most probably the large yellow-spotted many-flowered variety of the European L. *Pomponium*; of this any one that attends to its description may easily convince himself. Blooms in July and August; scentless; seeds freely and is easily propagated by the numerous offsets it produces; tolerably hardy; at least we never lost any in the severest winters by cold merely; the bulbs sometimes rot in very wet seasons. *G.*

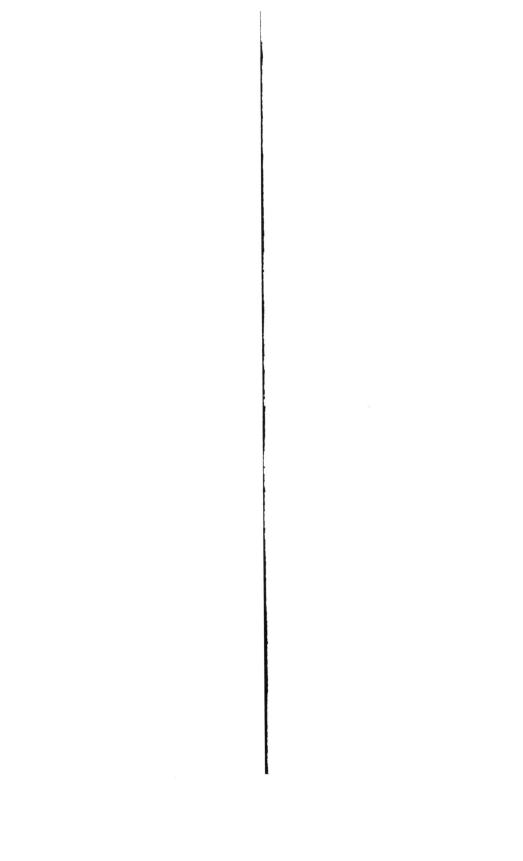

Syd. Edwards del.   Pub. by T. Curtis, St Geo Crescent, July 1 1806.   F. Sansom sculp

HYACINTHUS ORIENTALIS. GARDEN HYACINTH.

****

Claſs and Order.

HEXANDRIA MONOGYNIA.

Generic Charaƈer.

Cor. infera, tubuloſa, ſexfida, laciniis patentibus. *Stam.* tubo incluſa; filamenta fere tota adnata. *Stylus* triqueter; ſtigma depreſſum. *Capſ.* ovato-trigona. *Sem.* plura, ſubrotunda. G.

OBS. *Radix bulbus tunicatus teƈus induſiis ſcarioſo-membranaceis. Folia radicalia, ſubſucculenta, concava, varie lorata. Scapus teres, carnoſus.* In orientali *e ſummo germine per trinos haud manifeſtos poros tres mellea exſudantur guttulæ; hinc optime dignoſcenda eſt ſpecies.* G.

Specific Charaƈer and Synonyms.

HYACINTHUS *orientalis* racemo ſparſim multifloro; bracteis minutis; corolla deorſum cylindrica baſi ventricoſa, laciniis recurvo patentibus; ſummo germine tribus cryſtallinis gemmato guttulis. G.

HYACINTHUS *orientalis. Linn. Sp. Pl.* 454. *Cluſ. Hiſt.* 1. 174, 175. *Hort. Eyſt. Vern. ord.* 2. *fol.* 4, 5, 6, 7, 14, 15. *Mill. Ic. t.* 148. *Diƈ. n.* 6. *Willd. Sp. Pl.* 2. 167. *Desf. Flor. Atl.* 1. 307.

HYACINTHUS corollis, &c. *Gronov. Orient. n.* 115. 44.

ZUMBUL Indi. *Park. Par. t.* 121. *f.* 2. HYACINTHUS orientalis vulg. *Id. l. c. f.* 3, 5, 6.

This common ornament of our gardens is a native of the Levant, growing abundantly about Aleppo and Bagdad; DESFONTAINES met with it alſo on the coaſt of Barbary. It was

cultivated

cultivated here by GERARDE in 1596. Such as wifh to be informed minutely of the hiftory of this old favourite of the Florift, have but to confult a treatife entitled " DES JACIN-THES," publifhed by the late Marquis de ST. SIMON at Am-fterdam; but perhaps MADDOCK's Florift's Directory, or MILLER's Dictionary, may anfwer all ufeful purpofes as well. Amidft the rage for Tulips in Holland, this flower alfo came in for no fmall fhare of eftimation, from one to two hundred pounds fterling having been given for a fingle root of fome certain favourite variety.

Varies with double and femidouble, with white, red, blue, and yellow flowers; in fcent without end. The Harlem Gardeners diftinguifh two thoufand varieties by name; and acres are employed in the environs of that city for the culti-vation of thefe flowers; from thence we receive annually the beft bulbs. G.

Syd Edwards del. Pub by T. Curtis, St Geo. Crescent July 1 1806. F.S. sc

CROCUS SULPHUREUS (α). WORST YELLOW OR OLD CLOTH OF GOLD CROCUS.

****

Class and Order.

HEXANDRIA MONOGYNIA.

Generic Character.—Vid. N⁰ˢ. 845 & 860.

Specific Character and Synonyms.

CROCUS *fulphureus* bulbo-tubere tunicis membranaceis brunneis tenuibus fibrofo-ftriatulis tecto; corolla æqualiter patente; antheris parvis, fagittatis, pallidis; ftigmatibus inæqualibus has longius fuper-antibus. *G.*

(α) corolla extus tribus lineis fufcis plumofis longitudinaliter percurfa. *G.*

CROCUS vernus flavus ftriatus. *Park. Par.* 163. *f.* 16.

CROCUS vernus latifolius flavo-vario flore. *Raii Hift. p.* 1174. *n.* 8. defcr. optima.

(β) corolla concolor; fulphurea abfque omni ftria aut macula. *G.*

CROCUS vernus latifolius flavo flore minore et pallidiore. *Bauh. Pin.* 66. *Tournef. Inft.* 352.

Narrow-leaved Spring CROCUS with fmaller Brimftone-coloured flowers. *Mill. Dict. ed.* 7.

We believe this to be really a diftinct fpecies; at the fame time that we have no doubt that the fterility and fmallnefs of the anthers is not natural, but a mere degeneration produced by long culture in a climate differing from the native one. This appearance however they have retained at leaft fince the time of RAY, as his excellent defcription plainly fhews. We cannot believe it to be a variety of *mæfiacus*, from which it differs in the texture of the tunics and fize of the bulb-tubers, as alfo in fize and colour of corolla and pro-

portionate

portionate elevation of the stigmas; nor of *sufianus*, which has a very distinct bulb-tuber and the outer segments of the corolla revolutely patent. It appears to us to come nearest to *biflorus*, but has still very different kind of tunics to the bulb-tuber, which are much thinner striated and pliant, not even imbricated and subputamineous as in that; it differs also something in the organs, and entirely in colour. Varies with striped and plain flowers; both varieties being true to their specific characteristics. Never feeds, which *sufianus*, *biflorus* and *vernus* do abundantly, but *mæsiacus* more sparingly. Propagates most profusely by offsets; has no scent; blooms one of the earliest; is the least ornamental of any.

Both varieties were communicated by Mr. WILLIAMS, of Turnham-Green, than whom there is no more curious cultivator of this genus, as well as of almost the whole bulbous tribe of plants. G.

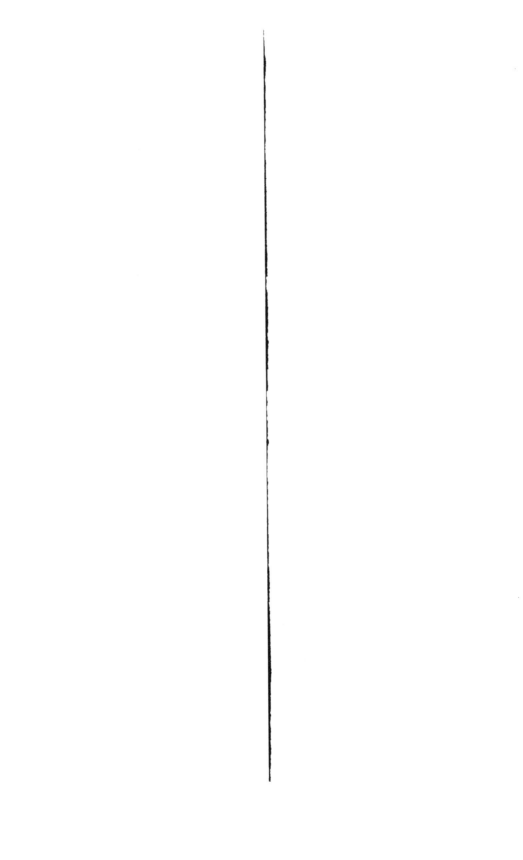

Edwards del. Pub. by W Curtis St Geo Crescent July 1 1806. F Sansom sculp.

Generic Character. *Vide N^{um}·* 919.

Specific Character and Synonyms.

SCILLA *romana* foliis fcapo longioribus attenuatis convoluto-concavis ; racemo confertiore cylindrico-conico ; braĉteis minimis, fubcalcaratis, craffis ; pedicellis corollæ æqualibus; hacce cyathiformi-campanulata ultra medium partita ; filamentis ufque bafin laciniarum liberis, planis, corollæ fubæqualibus. G.

HYACINTHUS *romanus. Linn. Syft.* 335. *Mant.* 224. *Hort. Kew.* 1. *p.* 458. *Mart. Mill. Diĉt. Willd. Sp. Pl.* 2. 169. *Desf. Fl. Atl.* 1. 308.

HYACINTHUS comofus albus belgicus. *Baub.* 42.

HYACINTHUS comofus albo flore. *Cluf. Hift.* 1. 180.

HYACINTHUS comofus byzantinus. *Hort. Eyft. Vern. ord.* 2. *fol.* 11. *f.* 2.

HYACINTHUS comofus albus cum cæruleis ftaminibus. *Baub. Hift.* 2. 584.

HYACINTHUS comofus. *Park. Par. t.* 117. *f.* 1.

While we adhere to the three very artificial, yet convenient, feĉtions of SCILLA, HYACINTHUS, and MUSCARI, this muft be referred to the former, on account of the far-parted corolla and free ftamens; although from appearance we fhould rather have ranked it under the latter. LINNÆUS tells us it grows in and about the city of Rome; DESFONTAINES found it on the coaft of Barbary; Mr. LAMBERT has a fpecimen brought from Tangiers. Said in *Hortus Kewenfis* to have been introduced by Mr. GRÆFER about 1786; it is however recorded both by PARKINSON and GERARDE. Hardy.

Our

Our drawing was made at Mr. MALCOLM's Nurfery at Kenfington.

Bulb ovate, about the fize of a hen's egg, covered with brown fcariofely membranous tunics. Leaves 4—5, far longer than fcape, from a broader bafe attenuated, convolute-concave, ftreaked, ambient. Scape round upright. Raceme clofifh, cylindro-conical. Pedicles about the length of the corolla. Bractes minute, glandularly thickened, fpurred, becoming gradually obfolete towards the top of the raceme. Corolla blue-white, fubcyathiformly campanulate, parted three-fourths of the length, corrugate outwards, patulous upwards, nodofely angular at the bafe, outer fegments thickened at the top and protuberant outwards. Filaments white, nearly equal to the corolla, linear, flat, contiguous, equal, adhering to corolla at bafe of the fegments; anthers fmall, blue, fagittate. Germen oval, obtufely alate, fubtrilobate, fix-ftreaked; ftyle trigonally briftleform; Stigma depreffed, fimple, blueifh; Capfule trilobately alate, lobes attenuately compreffed nearly as in MUSCARI. Flowers without fcent. Blooms in May. Seeds freely. Rather a fcarce plant in our gardens. G.

ERRATA.

No. 934. p. alt. l. 38. ante " R." adde " H."
No. 935. l. 20, dele " Lufit."

S.J Edwards del   Pub. by T Curtis S¹ G⁰ Crefoot July 1 1806   F Sansom sculp

NARCISSUS ORIENTALIS (a). NARCISSUS OF THE LEVANT.

Class and Order.

HEXANDRIA MONOGYNIA.

Generic Character.—Vid. N^{u.} 924 & 925.

Specific Character and Synonyms.

NARCISSUS *orientalis (stamina inæqualia alterna breviora tubo inclusa adnata)* 2-multiflorus; foliis latioribus, loratis, parum concavis; scapo subtereti; corollæ laciniis deflexo-patentibus, ellipticolanceolatis; corona his triplo breviore, subrotato-cupellata, subplicato-rugosa, trilobo-fissa, eroso-crenulata; tubo limbum subæquante. *G.*

NARCISSUS *orientalis. Mant. 62. Hort. Kew. 1. p. 409. Mart. Mill. Dict. Willd. Sp. Pl. 2. 58.*

NARCISSUS *Gron. orient. n. 99. p. 38.*

NARCISSUS latifolius byzantinus medio luteus II. *Cluf. Hift.* 1 154.

NARCISSUS narbonensis major amplo flore—medio croceo polyanthos—narbonensis medio luteo serotinus major. *Park. Par.* 82.

NARCISSUS polyanthos orientalis calyce medio luteo odoratus maximus. *Hort. Eyft. Vern. ord. 3. fol.* 15. *f.* 1, 5.

NARCISSUS orientalis medio croceo major—constantinopolitanus minor calyce fimbriato medio croceo. *Eod. l. c. fol.* 12. *f.* 2, 3.

BASELMAN major. *Trew. Seligm. 1. t.* 23.

(α) 2—4 florus; corona crocea; laciniis albidis. *G.*

(β) 2—4 florus; laciniis pallide luteis; corona lutea profundius trilobatim fissa. *G.*

(γ) 6—multiflorus; laciniis niveis, corona sulphurea aut citrina. *G.*

We

We have omitted all the figures from RUDBECK cited by LINNÆUS for synonyms, as they appear to us to belong exclusively to *biflorus*, from which our plant may be distinguished by its proportionately longer and trilobately cleft crown, as well as by not having the edges of the outer leaves turned up. Are they however really distinct species? (γ) is the variety of *Hort. Kew.* Differs from *Tazetta*, to which it is also but too closely akin, by a crown more manifestly trilobate, more crenulate and patent. All the varieties of our present species are very fragrant and bloom earlier than either *poeticus* or *biflorus*, but later than *Tazetta*. In the ensuing fasciculus we mean to give (β) and (γ). The best bulbs of this species are imported from Holland. Hardy.

According to the appellation bestowed on it by the older Botanists, a native of the Levant; probably some of the varieties are also found in Spain and the South of France. Not figured as a Linnean species in any work known to us. *G.*

Pub by T Curtis St Geo Crefant July 1.1806   F Sansom sculp

CLAYTONIA VIRGINICA. VIRGINIAN CLAYTONIA.

✚✚✚✚✚✚✚✚✚✚✚✚✚✚✚✚✚✚✚✚✚

Clafs and Order.

PENTANDRIA MONOGYNIA.

Generic Charatter.

Cal. bivalvis. *Cor.* 5-petala. *Stigma* 3-fidum. *Capf.* 3-valvis, 1-locularis, 3-fperma.

Specific Charatter and Synonyms.

CLAYTONIA *virginica;* foliis lineari-lanceolatis, petalis integris. *Hort. Kew.* 1. *p.* 284. *Willd. Sp. Pl.* 1. *p.* 1185. *Mart. Mill. Diff. a.* 1.

CLAYTONIA *virginica. Sp. Pl.* 294. *Reich.* 1. *p.* 572. *Decand. plant. gr.* 131. *Michaux Flor. Bor. Am.* 1. *p.* 160.

ORNITHOGALO affinis virginiana, flore purpureo pentapetaloide. *Pluk. Alm.* 272. *t.* 102. *f.* 3. *Rudb. Elyf.* 2. *p.* 139. *f.* 6.

The variety with broader lanceolate leaves mentioned in Hortus Kewenfis, is probably the *caroliniana* of MICHAUX. It occurs alfo with flowers of a deeper rofe colour : in our plant the petals are white ftreaked with red veins. JUSSIEU has placed CLAYTONIA in his natural order of *Portulaceæ* together with MONTIA, to which it has certainly a very near affinity. If CLAYTON's obfervation be correct, that the feed is monocotyledonous, perhaps it fhould be brought nearer to the *afphodeli,* to which family it approaches in general habit, in having a tuberous root, a fcape in part embraced by the leaves, which are not always exactly oppofite, a two-valved perfiftent calyx in fome refpects refembling a fpathe, a corolla decaying before it falls off, a trifid ftigma, and a three-valved capfule. This is however one-celled, and contains three kidney-fhaped feeds, or rather lentiform, with a notch at the part from whence the umbilical cord iffues, by means of which it is connected with the bottom of the capfule. The embryo of the feed is rolled round a farinaceous perifperm.

A native of moift woods in Virginia and New-England. Communicated by Mr. WILLIAMS of Turnham-Green, and Mr. SAMUEL CURTIS of Walworth. A hardy perennial. Flowers in May. Propagated by feeds or by the tuberous roots. Requires a moift foil in a fhady fituation.

Introduced by Mr. J. CLAYTON before 1759.

Syd. Edwards del. Pub. by T. Curtis. St Geo Crescent July 1 1805. F. Sansom sculp

PRIMULA INTEGRIFOLIA. ENTIRE-LEAVED PRIMROSE.

✹✹✹✹✹✹✹✹✹✹✹✹✹✹✹✹✹✹

Clafs and Order.

PENTANDRIA MONOGYNIA.

Generic Charaĉter.

Involucr. umbellulæ. *Corollæ* tubus cylindricus : ore patulo.

Specific Charaĉter and Synonyms.

PRIMULA *integrifolia ;* foliis integerrimis ellipticis, ad oras fubcrenato-cartilagineis, umbella ereĉta, calycibus longe tubulofis obtufiffimis. *Jacq. Mifc.* 1. *p.* 160. *Willd. Sp. Pl.* 1. *p.* 805. *Mart. Mill. Diĉt. a.* 16.

PRIMULA *integrifolia. Sp. Pl.* 205. *Jacq. Vind.* 209. *Obf.* 1. *p.* 26. *t.* 15. *Fl. Auft. t.* 327. *Scop. Carn. n.* 208. *Allion. Ped.* 1. *p.* 93 ?

PRIMULA foliis ellipticis carnofis integerrimis. *Hall. Helv. ? n.* 615.

PRIMULA *incifa. Lamarck Fl. Franc.* 2. *p.* 250 ?

SANICULA alpina rubefcens folio non ferrato. *Bauh. Pin.* 243.

AURICULA urfi carnei coloris foliis minime ferratis. *Bauh. Hift.* 3. *p.* 868.

AURICULA urfi quarta. *Cluf. Hift.* 1. 304. *Ejufd. Pann.* 349.

DESCR. *Root* perennial. *Leaves* growing thick together, oblong-elliptic, dilated at the bafe, flefhy, rigid, quite entire, with a very narrow white cartilaginous margin, fhining on the upper furface, whitifh on the under. *Scape* fhorter than the leaves, bearing about three purple flowers with a white centre in an umbel. *Braĉtes* one to each flower, linear and longer than the pedicle. *Calyx* cylindrical, longer than the pedicle, coloured at the upper part, five-toothed ; teeth ereĉt, obtufe, frequently emarginate. *Corolla* funnel-fhaped : tube longer than the calyx, fwollen in the middle and expanded upwards : limb patent, five-cleft : laciniæ obcordate, veined. *Filaments* red, fhort, inferted into the tube : anthers ereĉt-incumbent : pollen deep yellow. *Ovary* globofe, ftyle half the length of the tube of the corolla ; ftigma capitate.

We doubt whether the fynonyms from HALLER, LAMARCK, and ALLIONI, belong to our plant, which was raifed by Mr. LODDIGES from feeds fent him from Auftria feveral years ago, and is certainly the one defcribed by JACQUIN and long before by CLUSIUS. A hardy perennial, increafing rapidly by offsets from the roots, but very rarely flowering with us.

S.J. Edwards del. Pub by T. Curtis St Geo Crescent July 1 1806 F. Sansom sculp

Convolvulus Bryoniæ-Folius. Bryony-Leaved Bindweed.

✿✿✿✿✿✿✿✿✿✿✿✿✿✿✿✿✿✿✿

Class and Order.

Pentandria Monogynia.

Generic Character.

Cor. campanulata, plicata. *Stigm.* 2. *Capf.* 2-locularis: lo-culis difpermis.

Specific Character and Synonyms.

CONVOLVULUS *bryoniæ-folius;* foliis feptemlobo-palmatis hifpidis: lobo medio finuato produɕo, pedunculis axillaribus folitariis longiffimis articulatis.

Descr. *Stem* twining, herbaceous, hifpid. *Leaves* hifpid on both fides, varying in fhape, upper ones generally divided into feven unequal lobes, of which the middle one is much the largeft, finuated, and pointed: lower leaves near the foot more entire, oblong-cordate, irregularly finuated. *Petioles* nearly the length of the leaf, channelled on the upper fide. *Peduncles* growing fingly from the axils of the leaves, often twice the length of both leaf and petiole, jointed and frequently branching at the joints, bearing from one to three flowers; at the upper part of the plant the peduncles are generally fhorter and only one-flowered. *Bractes* two, fmall, fubulate, oppo-fite at each joint. *Calyx* 5-leaved; *leaflets* ovate, margined, preffed clofe together. *Corolla* fhewy, large, reddifh purple, ftriped: margin nearly entire with five fmall teeth. *Filaments* fubulate, half the length of the corolla, inferted at the bafe of the corolla. *Anthers* fomewhat arrow-fhaped. *Germen* fupe-rior, ovate, fmooth, two-celled. *Style* ereɕ, equal to the filaments. *Stigmas* two.

It is too nearly allied to Convolvulus *althæoides* (No. 359) but is a much more robuft plant; has no filkinefs or filvery whitnefs in the leaves; the flowers are larger and deeper coloured.

Introduced by Isaac Swainson, Efq. who raifed it from feeds received from China in 1802; the young plants were preferved in the ftove through the firft winter, and planted in the open border in the fpring. Mr. Swainson confiders it as a hardy perennial, thriving beft in a fouth border. Flowers from June to Auguft, and perfeɕs its feeds in the autumn. Our drawing was made at the Botanic Garden in Brompton.

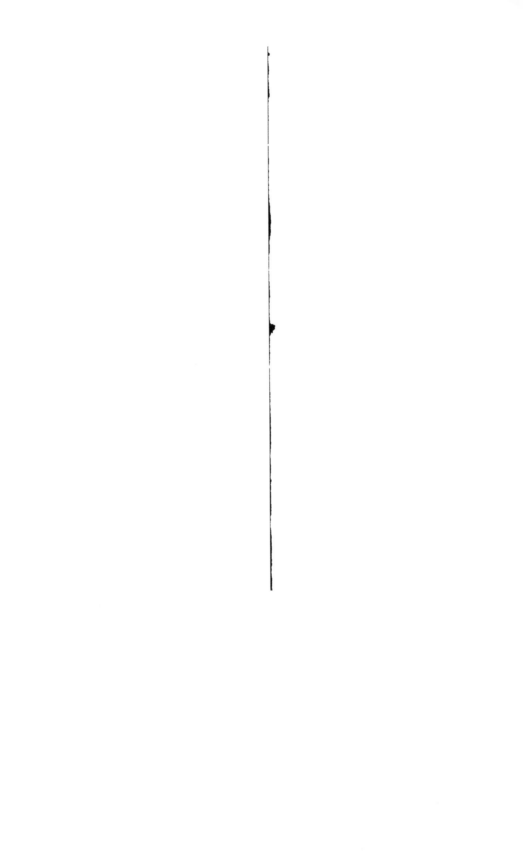

Pub. by T Curtis, St Geo Crescent July 1 1806. F Sansom

Dillwynia Glaberrima. Smooth-Leaved Dillwynia.

✳✳✳✳✳✳✳✳✳✳✳✳✳✳✳✳✳✳

Clafs and Order.

Decandria Monogynia.

Generic Charaɛter.

Cal. fimplex 5-fidus, 2-labiatus. *Cor.* papilionacea. *Stylus* reflexus. *Stigma* obtufum, pubefcens. *Leg.* ventricofum, 1-loculare, 2-fpermum. Smith.

Specific Charaɛter and Synonyms.

DILLWYNIA *glaberrima ;* foliis lævibus, floribus terminalibus fubcapitatis. *Smith in Ann. of Bot. v. 1. p.* 510.

Descr. *Stem* fhrubby, ereɛt, branched, hairy, rough, and as it were fluted with the permanent woody part of the old petioles. *Leaves* fimple, linear, rigid, fmooth, terminated in an oblique fubpungent mucro, thick-fet, patent, on fhort ad-preffed petioles inferted by a joint into a permanent woody theca, extending a little down the ftem. *Inflorefcence* a terminal capitulum of about fix flowers on very fhort peduncles with one minute braɛte. *Calyx* fomewhat coloured, perfiftent, nearly bell-fhaped, five-cleft: fegments nearly equal, fubbilabiately arranged, the two upper ones being fomewhat largeft and nearer together, the three lower more diftant. *Corolla* papilionaceous, bright yellow with a red ftarry fpot in the centre: vexillum two-lobed, with a claw narrow, channelled, and con-traɛted at the upper part. *Alæ* ftraight, half the length of the vexillum, and twice as long as the carina which is very fmall, adhering at the point, where it is of a red colour, diftinɛt to-wards the claws. *Stamens* ten: filaments conneɛted at the bafe, but for the moft part free: anthers yellow, fimple. *Ovary* ob-long, villous: ftyle bent back: ftigma truncated, appearing to us moiftened with a fine dew, but not villous.

This genus, of which two other fpecies have been figured in the Exotic Botany, was named by Dr. Smith in honour of Mr. Lewis Weston Dillwyn, author of a very accurate work on Englifh Confervæ.

Our drawing was made from a plant communicated by Mr. Loddiges of Hackney. Is a native of New-Holland. Re-quires to be proteɛted from froft by a greenhoufe and a treat-ment fimilar to that fuitable to Pultenæa, Platylobium, &c.

Propagated by feeds, which we have not as yet feen pro-duced with us.

Syd. Edwards del. Pub. by Curtis, St Geo Crescent Aug 1 1806   F Sansom sculp.

NARCISSUS TRILOBUS. NARROW-LEAVED NARCISSUS.

✱✱✱✱✱✱✱✱✱✱✱✱✱✱✱✱

Clafs and Order.

HEXANDRIA MONOGYNIA.

Generic Character.—Vid. N°s. 924 & 925.

Specific Character and Synonyms.

NARCISSUS *trilobus (ftamina alterna longiora de tubi ore emi-cantia)* fubtriflorus ; foliis anguftioribus cana-liculato-femiteretibus ; fcapo tereti ; corolla geniculato-nutante ; laciniis ftellatis tubo fub-æqualibus ; corona grandiufcula fubcylindrico-cupulata, levius trifida, integriore, his fubduplo breviore ; ftylo ultra hanc exferto. G.

NARCISSUS *trilobus. Sp. Pl.* 415. *Syft. Veg. Murr.* 317. *Willd. Sp. Pl.* 2 38. *Mart. Mill. Dict.*

NARCISSUS *nutans. Haworth Mifc. Nat.* 179.

NARCISSUS *juncifolius latiore calyce. Hort. Eyft. Vern. Ord.* 3. *fol.* 14. *f.* 2.

NARCISSUS *anguftifolius pallidus calyce flavo. Baub. Pin.* 51. *Rudb. Elyf.* 2. *p.* 61. *f.* 3.

For the fpecimen from which our drawing was made we have to thank Mr. HAWORTH, who fent it us under the name of NARCISSUS *nutans.* We do not know of any publi-cation in which this has been figured as a Linnean fpecies ; nor is it enumerated among thofe of *Hortus Kewenfis.* Said to be a native of the South of Europe. The fingle fpecimen, of which a figure is here publifhed, is the only one we have ever feen. It feems to be intermediate between *jonquilla* and *triandrus,* having the crown longer than the firft and fhorter than the other. The germen is oblong oval, trigonal, and
large.

large.   Blooms in May. : *Trilobus* is rather an unlucky name for it, as others of the genus are more conspicuously trilobate.

Since publishing the N<small>ARCISSUS</small> *mofchatus* (No. 924) we have seen the above-quoted work of Mr. H<small>AWORTH</small>, and find that his N<small>ARCISSUS</small> *tortuofus* may be added as a synonym to that Linnean species; the flower of which, in an earlier stage, is far more cernuous than in our drawing of it.   *G.*

Syd Edwards del   Pub by T.Curtis St Geo Crescent Aug 1

NARCISSUS ORIENTALIS (γ). MANY-FLOWERED NARCISSUS OF THE LEVANT.

****

Clafs and Order.

HEXANDRIA MONOGYNIA.

Generic Charaĉter.—Vid. N$^{os.}$ 924 & 925.

Specific Charaĉter and Synonyms.

NARCISSUS *orientalis (Vid. N$^{os.}$ 940).*
(γ) 6-multiflorus; laciniis niveis corona fulphurea. *G.*
NARCISSUS *Tazeta* (bicolor). *Lil. a Redouté, p. 17. abf-
que ic.*
NARCISSUS medio luteus copiofo flore odore gravi. *Rudb.
Elyf.* 2. *p.* 57. *f.* 11.

This and our *papyraceus* were, we have no doubt, included by LINNÆUS in his *Tazeta*, but to us they appear fufficiently diftinĉt, and this an undoubted variety of *orientalis ;* in fome of the Dutch catalogues it is known by the name of the *Grande Primo Citroniere.* G.

Fd. Edwards del. Pub. by T Curtis St Geo Crescent Aug 1 1806 F. Sansom sculp

NARCISSUS PAPYRACEUS (α). ITALIAN OR PAPER-WHITE NARCISSUS.

✳✳✳✳✳✳✳✳✳✳✳✳✳✳✳✳✳

Class and Order.

HEXANDRIA MONOGYNIA.

Generic Character.—Vid. Nᵒˢ· 924 & 925.

Specific Character and Synonyms.

NARCISSUS *papyraceus (stamina adnata tria breviora intra tubum inclusa)* multiflorus; foliis lorato-concavis; scapo ancipiti, subplano compresso, striato; laciniis stellatis tubo subæqualibus; corona cupulata, his 3-4plo breviore, crenulato-erosa; stylo intra coronam. *G.*

NARCISSUS *Tazeta* (albus). *Redouté Lil. p. 17. absque ic.*

NARCISSUS *Tazeta. Linn. Sp. Pl.* 416. *Hort. Kew.* 1. *p.* 410. *Willd. Sp. Pl.* 2. 39.

NARCISSUS pisanus vel totus albus. *Park. Parad.* 81. *f.* 4.

NARCISSUS latifolius flore prorsus albo. *Floril. Auct.*

NARCISSUS latifolius simplici flore prorsus albo. 1, 2. *Cluf. Hist.* 1. 155.

(α) multiflorus; corolla tota alba; stylo parum ultra tubum porrecto; foliis glaucis. *G.*

(β) 4—6-florus; laciniis ochroleucis, corona pallide sulphurea; stylo coronam subæquante; foliis haud ita glaucis. *G.*

ROMAN NARCISSUS.

Very probably this, or the variety γ of NARCISSUS *orientalis,* may have been the plant designed by LINNÆUS for the type of his *Tazeta,* as likely indeed as the one we have given under that title; but, as he has evidently combined more than one species in his synonymy, we have thought it most useful to apply his name to the one which had been already figured

under

under it in REDOUTE's work, and to adopt another for this plant, which we think diſtinᚇ. Our ſpecies is poſſibly the *crenulatus* of Mr. HAWORTH, but his charaᚇer is too ſhort and indefinite to enable us to determine this faᚇ with certainty.

Differs from *Tazeta* in having a ſhallower crown, with an eroſely crenulate margin, a very much flattened ancipital ſcape, a ſmaller bulb, and an entirely white or a cream-coloured corolla.

The bulbs of this plant are uſually imported by the owners of Italian warehouſes immediately from Italy. Very ornamental and fragrant, eſpecially (β) called in the ſhops the ROMAN NARCISSUS, which is often imported in a double ſtate. *G.*

G.J. Edwards del Pub by J. Walker London Aug 1 1806. F. Sansom sculp

NARCISSUS ORIENTALIS (β). CREAM-COLOURED NARCISSUS OF THE LEVANT.

****

Class and Order.

HEXANDRIA MONOGYNIA.

Generic Character.   *Vide* N°ˢ· 924 & 925.

Specific Character and Synonyms.

NARCISSUS *orientalis.* *Vid.* N°ⁿ· 940.

(β) 2—4-florus; laciniis pallide luteis; corona lutea profundius trilobatim fissa. *G.*

In plants that have been long cultivated in gardens, it will often be a question not readily decided, whether certain individuals are mere varieties, arising from the adventitious circumstances of culture, or originally distinct species, or hybrids deriving their origin from intermixture. It may throw some light upon these questions to observe,

1. That whilst the industry of Florists, by means of long culture in appropriate soil and under nice management, (in some cases offering a too profuse nutriment, in others subtracting the due proportion), can produce an almost endless variety of appearances in the individuals of the same species, especially in stature, colour, and multiplication of parts, yet amidst this numerous host, the scrutinising eye of the Botanist will find little or no change made in the essential characters; these mostly remain the same in all. For although, by a multiplication of the petals or other parts, the smaller and more essential organs are suffocated; yet these changes affect the generic more than the specific distinction. And notwithstanding the immense number of these artificial varieties, such is the tendency of nature to maintain a uniformity in the individuals of the same species, that a few years neglect is sufficient to reduce the thousand gaudy inhabitants of the Florist's border to the state of their original parent.

2. The same industrious spirit, when any particular flower happens to be in fashion, will seek far and near for closely-

related

related species of the same genus, and by submitting these to a suitable culture will occasion similar changes, in the endless variety of which the species may easily be confounded with the one before in cultivation.

3. This blending of different species will be still farther promoted by the accidental or purposely-contrived admixture of them, by fecundating one with the pollen of another. An offspring partaking of both parents is the consequence, and in some cases perhaps a permanent intermediate variety, scarcely to be distinguished from a really distinct species, may be thus produced. For we cannot go so far, in either the vegetable or animal kingdom, as to assert with some authors, that hybrids or mules are in every case steril. Mule birds, as we are assured by breeders, will frequently procreate, and the offspring of the wolf and the dog has been proved to be prolific; but we are not thence to conclude, as some have done, that the canary and gold-finch, the wolf and the dog are in reality the same species. We can see however that the confounding of different species by these mixed productions is very limited, in many cases confined to the individuals first produced, in others terminating perhaps with the next or third generation, and if a longer existence be allowed, we may infer a very great proximity between the parent plants. In vegetables indeed the duration may be longer from their power of propagation by other means than by seeds, but the increase obtained by offsets from the roots, cuttings, or layers, when the fostering care of man ceases, will shortly perish. Hence a very long-continued existence without change will often afford a strong presumption of a plant's being a real species.

Armed with such reflections, the Botanist may boldly enter the Florist's parterre, undismayed by the barbarous host of a Dutch catalogue. Here he will certainly find, that if the older botanical writers frequently raised varieties to the rank of species, the modern have sometimes confounded such as are really, and ever have been, distinct.

To enable us to decide in difficult cases, it becomes necessary to study varieties as well as species; and this must be our apology for admitting several of the former into a work, in the general plan of which they are excluded. Faithful representations and accurate descriptions, when recorded, cannot fail to establish the truth in the end. *S.*

Del. Edwd. del. Pub. by T. Curtis St Geo Crescent Aug 1 1806. F. Sansom sculp

AOTUS VILLOSA. VILLOUS AOTUS.

Class and Order.

DECANDRIA MONOGYNIA.

Generic Character.

Cal. 5-fidus, fimplex. *Cor.* papilionacea: alis vexillo brevioribus. *Stylus* filiformis. *Stigma* obtufum. *Leg.* 1-loculare,
2-fpermum. SMITH.

Specific Character and Synonyms.

AOTUS *villofa.* *Smith in Ann. of Bot. v.* 1. *p.* 504. *Dryander*
    ibid. vol. 2. *p.* 519.
PULTENÆA *villofa.* *Bot. Repof.* 309.
PULTENÆA *ericoides.* *Vent. Malmaif.* 35. *Poiret in Encyc.*
    Meth. p. 738.

Dr. SMITH has, in our opinion, very properly feparated this
from the genus PULTENÆA; from his PULTENÆA *villofa,*
with which it has been confounded in the Botanift's Repofitory,
it is in every refpect different.

To what is faid by the Doctor in the Annals of Botany, we
have nothing to add, except that the calyx, befides wanting
the appendages, is bilabiate, and has the two teeth of the upper
lip fhorter and more divaricate than the reft.

It is a pretty little greenhoufe fhrub; native of New-
Holland; flowers in May; communicated by Mr. LODDIGES
of Hackney.

Pub. by Curtis Glazen Orchard Aug.¹ 1806   P. Sanson sculp.

ARUM TRIPHYLLUM (α) ZEBRINUM. ZEBRA-FLOWERED ARUM.

✤✤✤✤✤✤✤✤✤✤✤✤✤✤✤✤✤

Clafs and Order.

MONŒCIA POLYANDRIA, *olim ad* GYNANDRIAM POLY-ANDRIAM *relatum.*

Generic Charačter.

Spatha monophylla, cucullata. *Spadix* fupra nudus, inferne fæmineus, medio ftamineus.

Specific Charačter and Synonyms.

ARUM *triphyllum;* acaule ; foliis trifoliolatis pedatis : foliolis ovalibus acuminatis, floribus dioicis.

ARUM *triphyllum. Sp. Pl.* 1368. *Willd. v.* 4. *p.* 480. *Gron. Virg.* 142. *Michaux Fl. Bor. Amer.* 2. *p.* 188.

α. *zebrinum ;* fpadice atro-purpureo fpatha vittata.

DRACUNCULUS feu Serpentaria triphylla Brafiliana. *Dodart. Mem.* 81. *cum fig. Robert. Ic. Raii Hift.* 1212. *certiffime. Bauh. Pin.* 195 ? *Prod.* 101 ?

β. *viride ;* fpadice fpathaque uniformiter viridefcentibus.

ARUM minus triphyllum feu Arifarum, pene viridi virginianum. *Morif. Hift.* 3. *p.* 547.

γ. *pallefcens;* fpadice furfum rubefcente, fpatha pallide vittata.

ARUM feu Arifarum minus marianum flore et pene ex pallido virefcente. *Pluk. Alm.* 39. *t.* 376. *f.* 3.

δ. *atropurpureum ;* fpadice fpathaque uniformiter atropurpureis.

ARUM *atro-rubens. Hort. Kew. v.* 3. *p.* 315. *Mart. Mill. Dičt. Willd. Sp. Pl. v.* 4. *p.* 481, *exclufo fynonymo Pluk.*

ARUM five ARISARUM triphyllum minus, pene atrorubente virginianum. *Pluk. Alm.* 52. *t.* 77. *f.* 5.

This variety, which is by far the moft beautiful, is alfo of the largeft growth, on which account we have placed it firft.

The

The ARUM *triphyllum* is said by LINNÆUS, who confidered the genus as gynandrous, to be monœcious, bearing male and female flowers from the fame root on different ftalks; we apprehend however that this is a miftake, and that it is really diœcious, the male and female flowers rifing from different roots; as has certainly been the cafe in fuch as we have feen. Our plant is female, bearing a number of germens, each furmounted by a white ftigma, crowded together at the bafe of the fpadix. Above thefe are a few fcattered, irregular, antherlike maffes, but which do not appear to contain pollen, and as the fame are found above the ftamens in the male plant, they may perhaps be confidered as nectaries.

The footftalk of the leaf and fcape are involved in two or three truncated fheaths of a pale colour fpotted with purple, the former exceeding the latter in length, fo that the leaf ftands over the flower, like an umbrella. In our plant the leaflets were not quite entire; the margins being fomewhat undulated and repand, which gives them the appearance of being flightly toothed, but they are not really fo. Native of North-America from Canada to Carolina. Probably the Brafilian plant may not be the fame fpecies. The roots are apt to perifh from the heat of the fun in dry weather unlefs planted in the fhade and a moift foil, and in fuch fituations they are liable to be deftroyed in the winter; it will be therefore fafeft to take them up after the leaves decay, and keep them in fand till the following fpring. Flowers in May. Communicated by Meffrs. NAPIER and CHANDLER, Vauxhall.

N.º 951

RHODODENDRON MAXIMUM. LAUREL-LEAVED RHODODENDRON.

✱✱✱✱✿✱✱✱✱✱✱✱✱✱✱

Clafs and Order.

DECANDRIA MONOGYNIA.

Generic Character.

Cal. 5-partitus. *Cor.* fubinfundibuliformis. *Stam.* declinata. *Capf.* 5-locularis.

Specific Character and Synonyms.

RHODODENDRON *maximum ;* foliis oblongis glabris fubtus difcoloribus margine acuto reflexo, umbellis terminalibus congeftis, corollæ laciniis concavis.

RHODODENDRON *maximum. Sp. Pl.* 563. *Willd. Arb.* 286. *Ejufd. Sp.* 2. *p.* 607. *Trew. Ebret. p.* 32. *t.* 66. *Wangenb. Amer. p.* 63. *t.* 23. *f.* 49. *Mart. Mill. Dict. n.* 9. *Michaux Fl. Am. Bor.* 1. *p.* 259. *Gærtn. Fruct. v.* 1. *p.* 304. *t.* 63.

KALMIA foliis lanceolato-ovatis nitidis fubtus ferrugineis, corymbis terminalibus. *Mill. Ic. t.* 229.

CHAMÆRHODODENDROS, lauri folio fempervirens, floribus bullatis corymbofis. *Catefb. Car.* 3. *p.* 17. *t.* 17. *f.* 2.

LEDUM lauro-cerafi folio. *Amœn. Acad.* 2. *p.* 201.

This elegant tree, according to CATESBY, adorns the weftern and remote parts of Pennfylvania, always growing in the moft fteril foil, or on the rocky declivities of hills and river banks, in fhady and moift places. MICHAUX fays it is

found

found from New-England to North-Carolina. In its native soil it attains the height of sixteen° feet.

There is much affinity between this tree and the RHODO-DENDRON *ponticum*, nor do the diftinguifhing chara&ers adopted by WILLDENOW feem fufficient; for the leaves of the latter are not of the fame colour on the under furface as the upper, and both are very fubje&, to vary in this refpe&, as well as in fize and form. The flowers of the *maximum* are paler coloured and grow in a much more compa& umbel upon fhorter pedicles; the corolla is more deeply divided and the fegments are rounder, more concave, and not undulated as in *ponticum*. The difficulty has of late been increafed by the raifing of hybrid varieties from an intermixture of the two. The upper fegment of the corolla, not the lower as is faid by WILLDENOW, is rather larger .than the reft, and beautifully fpotted towards the bafe with green and yellow.

Requires a moift foil with an admixture of bog-earth and a fhady fituation. Bears forcing as well as the *ponticum*, but is not fo well adapted to this purpofe from the palenefs of the flowers, which, in this ftate, become white, except the upper lacinia. Is rather more apt to be disfigured by the cold eafterly winds occurring late in the feafon.

Our drawing was made in the fpring of 1785, at Meffrs. WHITLEY and BRAME's, Old-Brompton, when every fhrub both of this and *ponticum* produced abundance of flowers in the greateft perfe&ion. This year not a complete umbel was to be feen in the whole colle&ion!

Introduced in 1736, by PETER COLLINSON, Efq.

Wh. mark del   Pub by T Curtis St Geo Crescent Sep 1806   F Sansom sculp

FRITILLARIA RACEMOSA. BUNCH-FLOWERING FRITILLARIA.

✸✸✸✸✸✸✸✸✸✸✸✸✸✸✸✸✸✸✸

Class and Order.

HEXANDRIA MONOGYNIA.

Generic Character.—Vid. N⁰ⁿ· 664.

Specific Character and Synonyms.

FRITILLARIA *racemosa;* racemo erecto 4—9-floro, foliofo, fubfecundo; foliis deorfum numerofis, fub-confertis, lineari-acuminatis, planis, glaucis; *Cætera* FRITILLARIÆ *Meleagridis.* G.

FRITILLARIA *pyrenaica. Sp. Pl.* 436. *Syst. Vegetab. Murr.* 325. *Hort. Upf.* 81. *Willd. Sp. Pl.* 2. 91. exclufis femper fynonymis *Cluf. app. Lob. adv. Park. Par.* 43. *f.* 11. atque *Bauh. Pin.* 64.

FRITILLARIA *pyrenaica* (β) *fupra No.* 664. rejectis fyno-nimis *Park. Parad.* 43. *f.* 12. *Swert. Flor.* 7. *f.* 2. cum eo *Bauh. Pin.* 64.

FRITILLARIA e foliorum alis florens (♂.) *ferotina atropur-purea. Hort. Cliff.* 119.

FRITILLARIA nigra floribus adfcendentibus. *Mill. Dict.* 3. rejecto fynonymo.

This is evidently the plant from which LINNÆUS charac-terized his *pyrenaica,* a name he adopted under the perfuafion that it was a variety of the true Pyrenean vegetable he found defcribed and figured in the works of CLUSIUS and LOBEL; in this error we followed him in the 664th number of this work; fince then we have obtained a living fpecimen, and can have no doubt of the fpecies being as diftinct from each other as any other two of the fame genus. In fact, if a variety of any known fpecies, it muft be of *Meleagris;* from which how-ever

ever it differs in the characters given above in our specific phrase. The corolla of *pyrenaica* is of a thick, coriaceous, fleshy substance, has a patulous margin, with the alternate segments twice as broad as the others and obovate; characters not to be found in our present subject, the leaves of which are also much more numerous, narrower, sharper, and more sparse. Its segments are rather less acute than those of *meleagris*, and its nectary is rather nearer the base of the segments than in that, but yet not so near as in *pyrenaica*; all three differ from *latifolia*, in having green, slenderer, and more patent stigmas. As LINNÆUS applied his specific title to this plant, under the idea of its being a variety of the one we have before published under that name, and which is really of Pyrenean origin, we have thought better to leave that appellation with it (especially as it was also included by LINNÆUS in his species) and adopt another for the present plant.

We are ignorant of its real habitat, possibly a mere variety of *Meleagris*. Blooms somewhat later than that or *pyrenaica*; quite scentless.

Our drawing was made from Mr. WILLIAMS's collection at Turnham-Green. Cultivated here in the time of MILLER.

Probably FRITILLARIA *bispanica umbellifera* of BAUHIN, PARKINSON and SWERTIUS, is really a variety of the *pyrenaica*; but this we have not yet met with. G.

Pub. by T. Curtis, St Geo Crescent Sep 1.1806. F. Sansom sculp.

ORNITHOGALUM UNIFOLIUM (β). GIBRALTAR STAR OF BETHLEHEM.

Clafs and Order.

HEXANDRIA MONOGYNIA.

Generic Character.—Vid. N°s. 653 & 746. Obf.

Specific Character and Synonyms.

ORNITHOGALUM *unifolium. Vid. fupra N^um. 935.*
(β) fpica fubcylindraceo, multifloro, conferto, floribus erectis,
 odoratis; foliis 2—3, breviter cufpidatis. *G.*
ORNITHOGALUM *concinnum. Salifb. Prod. Hort.* 240.
ORNITHOGALUM *nanum. Var.* 2. foliis tribus, fcapo
  unico in eodem bulbo. *Brot. Flor.*
  Lufit. 1. 250. ?

We cannot bring ourfelves to think this any other than a
variety of the above fpecies. This variety is faid to have been
found in the country near Gibraltar, whence it was received
in 1780, by the late Dr. FOTHERGILL. Flowers fweet-fcented.
Should be fheltered in a pit or garden frame.

Our drawing was made from a plant in the very felect
collection of Mr. WILLIAMS, Nurferyman, at Turnham-
Green. *G.*

Syd Edwards del Pub by Tuarlis S'Gre Crelcont Sep 1 1806 F Sansom sculp

Generic Character.—Vid. N°· 470.

Specific Character and Synonyms.

TRILLIUM *cernuum* flore pedunculato, cernuo. *Linn. Sp.*
Pl. 484. *Hort. Kew.* 1. 490. *Mill. Dict.* 1.
Smith Spicil. t. 4. *Michaux Flor. Bor-Amer.* 1.
216.

SOLANUM *triphyllum*, flore hexapetalo carneo. *Cat. Car.* 1.
45. *t.* 45.

PARIS foliis ternis, flore pedunculato nutante. *Cold. Noveb.*
1. 45.

After the detailed defcription in Dr. SMITH's SPICILE-
GIUM, we need not make any addition in this place, except
it be to remark, that a perfect trilocular fruit, fuch as reprefented in his figure, is in this genus at leaft dubious. The
receptacle of the feeds is in this fpecies formed by a projection
going off from the middle of three of the fides, but terminating
with a thickened extremity before it reaches the centre of the fruit.
In fuch a conftruction, of courfe, the fhrinking of the receptacles
of the feeds may occafion what appears to be a three-celled
ovary to become one-celled in the ripe fruit. Moreover a
difference in the length of the receptacle of the feeds in the
different fpecies of the fame genus, extending in one nearly or
quite to the centre of the fruit, in another lefs than half-way
towards

towards the centre, though evidently making no effential difference, will give in the former cafe the appearance of a three-
celled, in the latter that of a one-celled fruit. This obfervation will probably explain the feeming contradiction in the
formation of the fruit in this genus. Whether the three
feminal receptacles in any cafe perfectly unite in the centre,
as defcribed by Mr. SALISBURY in Paradifus Londinenfis,
No. 35, deferves to be further examined; in this fpecies they
are certainly free towards the centre and attached to the fides
of the fruit only.

A hardy plant, requiring fhade, and to be planted in bog-
earth. Found by MICHAUX in mountainous places in Upper
Carolina, by KALM in Canada, and by Mr. MENZIES in
Nova-Scotia. Cultivated here by MILLER.

Our drawing was taken from a plant communicated by
Meffrs. NAPIER and CHANDLER, Vauxhall.

Syd.Edwards del   Pub. by T.Curtis, St Geo:Crescent Sep.1.1806.   F.Sansom sculp.

*Generic Charaĉter.—Vid. N*ᵘᵐ· 916.*

Specific Charaĉter and Synonyms.

UVULARIA *perfoliata ;* foliis perfoliatis, ellipticis, obtuſis ;
corolla campanulata, intus ſcabrata ; antheris
cuſpidatis. *Exot. Bot. v.* 1. *p.* 97.
UVULARIA *perfoliata. Linn. Sp. Pl.* 437. *Mill. Diĉt.* 2.
Hort. Kew. 1. 434. *Willd. Sp. Plant.* 2. 94.
Michaux Flor. Bor-Amer. 1. 199.
(α) *major ;* calyce luteo. *Mich. l. c.*
POLYGONUM ramoſum flore luteo major. *Corn. Canad.*
38. *t.* 39.
(β) *minor ;* calyce pallide-exalbido. *Mich. l. c.*
UVULARIA *perfoliata. Exot. Bot. t.* 49.

Since we have not ſeen the living ſpecimen of this ſpecies,
we do not pretend to add to or alter what has been ſaid of the
plant in the *Exotic Botany.* According to the figures, our
plant ſeems to us to partake equally of Dr. SMITH's *flava* and
perfoliata. MICHAUX has two varieties, poſſibly his (α) is
the *flava* of Dr. SMITH. We ſtrongly ſuſpeĉt all theſe plants
will be found to be mere varieties of each other : (α) was found
by MICHAUX in Canada and on the very high mountains
of Carolina ; (β) in the mountains of middling height in Carolina and Virginia. MILLER ſays the ſpecies is perfeĉtly
hardy, and ſhould be planted in a hazel loam not too ſtiff
nor wet ; may be propagated by parting the roots about
Michaelmas, but not oftener than every third year. Blooms
in April and May.

Our drawing was made from a plant ſent us by Mr. WIL-
LIAMS, of Turnham-Green. G.

Edwards del   Pub by T. Curtis, St Geo Crescent Sep 1.1806   F. Sansom sculp

SCALIA JACEOIDES. KNAP-WEED SCALIA.

✳✳✳✳✳✳✳✳✳✳✳✳✳✳✳✳✳✳

Class and Order.

STNGENESIA POLYGAMIA SUPERFLUA.

Generic Character.

Receptaculum nudum. *Pappus* pilofus, fcaber, feffilis. *Corolla* radii infundibuliformes, irregulares.

Σκαλίας, *nomen a Theophrafto cuidam hujus ordinis planta adhibitum.*

SCALIA *jaceoides.*

DESCR. *Root* flefhy, tap-fhaped? perennial. *Stem* erect, round, fomewhat woolly, a little branched at the upper part. *Leaves* alternate, feffile, fpatulate-lanceolate, with rough margins, obfcurely three-ribbed; the midrib much ftouter than the lateral ones. *Flowers* uniformly yellow, folitary, on long, more or lefs fcaly peduncles. *Calyx* fubglobofe, imbricate, fcales linear, terminated with a broader, ovate-acuminate, fcariofe appendix. *Corolla* of the radius female, funnel-fhaped: tube filiform: limb fomewhat irregularly cut into from three to five narrow laciniæ. *Corolla* of the difk many, hermaphrodite, funnel-fhaped: tube filiform below. *Receptacle* naked, dotted, flat. *Seeds* oblong, rough, crowned with a feffile hairy fcabrous pappus, thofe of the difk and of the radius fimilar.

In habit, but not colour, this plant much refembles CENTAUREA *nigra*, and has the fame rigid rough afpect; but belonging to a different order in the Linnean fyftem, and having a naked receptacle, it will not unite in the fame genus with it. A native of New South-Wales, and boafts no great beauty, but as few fyngenefious plants have as yet found their way from that country into our gardens, it may be admitted for its rarity. May be treated as a hardy greenhoufe plant. Propagated by feeds, which however are not always perfected with us. Flowers in May and continues a long time in bloom. Introduced by Mr. LODDIGES of Hackney.

Syd Edwards del   Pub by T Curtis St Geo Crescent Sep 1 1806.   FSansom sculp

CAMPANULA ALPINA. ALPINE BELL-FLOWER.

❋❋❋❋❋❋❋❋❋❋❋❋❋❋❋❋❋❋❋

Clafs and Order.

PENTANDRIA MONOGYNIA.

Generic Chara&er.

Cor. campanulata, fundo claufo valvis ftaminiferis. *Stigma* 3-fidum. *Capf.* infera, poris lateralibus dehifcens.

Specific Chara&er and Synonyms.

CAMPANULA *alpina ;* caulo fimplici pedunculis unifloris axillaribus diphyllis. *Jacq. Vind.* 210. *Jacq. Auftr. 2. t.* 118. *Sp. Pl.* 1669. *Willd. Sp. Pl. 1.* 909. *Reich.* 463. *Mart. Mill. Di&. a.* 37.

CAMPANULA foliis ellipticis hirfutis, petiolis alaribus unifloris, floribus glabris. *Hall. Helv. n.* 695.

CAMPANULA alpina pumila lanuginofa. *Bauh. Pin.* 94.

TRACHELIUM pumilum alpinum. *Cluf. Hift.* 171. *Ejufd. Pann.* 687. *Park. Herb.* 645. 9. *Raii Hift.* 736. 21.

A native of the Alps of Switzerland and Schneberg in Auftria. The reflexed angles between the erect fegments of the calyx are fo very fhort, that it may admit of doubt whether it fhould have been arranged in the third or in the firft feFtion of this extenfive genus. The capfule is trilocular. It varies with pale afh-coloured flowers, and bright blue, inclined to violet. Is a hardy perennial, but requires the fame careful treatment as other alpine plants. Introduced from Auftria by Mr. LODDIGES, by whom it was communicated to us in flower at the latter-end of April. It continued for fome weeks in high beauty, and is indeed a very ornamental little plant.

Syd Edwards del   Pub by T Curtis, S.Geo Crescent Sep 1806   F Sansom sculp

Cal. Lab. fuperius abbreviatum, 2-dentatum ; inferius tridentatum, produ&ius. *Carina* corollæ truncata. *Legum.* pedicellatum, complanatum, dorfo gibbum, 2-fpermum. *Stigma* capitatum.

GOODIA *lotifolia.* *Salifb.* in *Parad.* *Lond.* 41.

Our drawing of this plant was taken at the Botanic Garden, Brompton, more than two years ago. We had before feen and defcribed it in Mr. HIBBERT's colle&ion at Clapham-Common, but were not able to fatisfy ourfelves perfe&ly refpe&ing its native country ; fufpe&ing from its habit, fo very different from that of moft of the Leguminofæ from New-Holland, that it was in reality a produ&ion of the Cape of Good Hope. We are informed however in the Paradifus Londinenfis, that it was found in New South-Wales, by PETER GOOD, and feeds of it tranfmitted by him to the Royal Garden at Kew.

This induftrious Gardener was induced by his love of plants to leave a lucrative employment and repair to fo diftant a country to colle& feeds for his Majefty ; in which fervice he died. By naming this plant after him, Mr. SALISBURY has endeavoured to perpetuate his memory, a duty which, we underftand, Mr. BROWN, fince his return from New South-Wales, had intimated his intention of fulfilling.

It is a hardy greenhoufe fhrub of handfome growth. Flowers in May, June, and July. Propagated by cuttings and feeds.

Syd. Edwards del. Pub by T Curtis S.t Geo Crescent Sep.t 1806 F Sansom sculp

CLEMATIS CALYCINA. MINORCA VIRGIN'S-BOWER.

Class and Order.

POLYANDRIA POLYGYNIA.

Generic Character.

Cal. o. *Petala* 4—6. *Semina* caudata.

Specific Character and Synonyms.

CLEMATIS *calycina ;* involucro calycino approximato, foliis ternatis, intermedio tripartito. *Hort. Kew.* 2. *p.* 259. *Vabl. Symb.* 3. *p.* 75. *L'Herit. Stirp. Nov.* 2. *t.* 26. *ined. Willd. Sp. Pl.* 2. *p.* 1289. *Mart. Mill. Dict. a.* 16.

CLEMATIS *balearica ;* scandens, foliis compositis tenuiter laciniatis, floribus calyculatis lateralibus, petalis interne guttatis. *Lamarck Encycl.* 2. *p.* 44.

Our drawing of this rare species of CLEMATIS, a native of the Island of Minorca, was taken at Mr. MALCOLM's nursery at Kensington. It requires the protection of a good greenhouse. Produces its flowers in the winter. Propagated by layers with difficulty. Introduced to the Royal Garden at Kew, in 1783, by M. THOUIN.

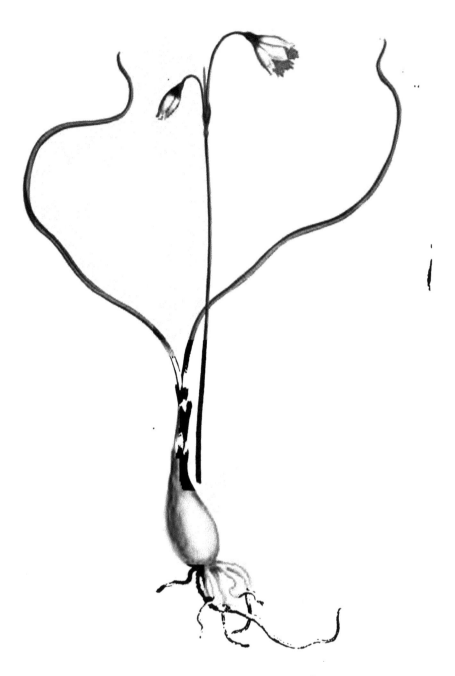

Syd. Edwards del Pub by T Curtis St Geo Crescent Oct 1. 1806   F Sansom sculp

LEUCOJUM AUTUMNALE. AUTUMNAL SNOW-FLAKE.

✳✳✳✳✳✳✳✳✳✳✳✳✳✳✳✳

Clafs and Order.

HEXANDRIA MONOGYNIA.

Generic Charaƈer.

Cor. campaniformis, 6-partita, apicibus, incraffata. *Stigma* fimplex.

Specific Charaƈer and Synonyms.

LEUCOJUM *autumnale ;* fpatha fub-biflora : laciniis triden-tatis, ftylis foliifque filiformibus.

LEUCOJUM *autumnale. Spec. Pl.* 414. *Willd.* 2. *p.* 30. *Reich.* 2. *p.* 16. *Læfl. It.* 136. *Hort. Kew.* 1. *p.* 406. *Brot. Fl. Luf. v.* 1. *p.* 552. *Desfont. Atl.* 1. *p.* 281. *Porret Voy. v.* 2. *p.* 144. *Parad. Lond.* 21. *Ic. Opt.*

LEUCOJUM bulbofum autumnale. *Baub. Pin.* 56. *Cluf. Hifp.* 271. *f.* 272. *Dod. Purg.* 410. *Pempt.* 230. *f.* 4. *Park. Parad.* 110. 2. *t.* 107. *f.* 10. *Raii Hift.* 1145. *Ger. Emac.* 148. *f.* 5. *Baub. Hift.* 1. *p.* 593. *fig.* 1. *Beft. Hort. Eyft. Ord.* 3. *aut.*

LEUCOJUM bulbofum tenuifolium minus flore rubello. *Grifl. Virid. Luf. n.* 1573.

TRICOPHYLLUM. *Renealm. Spec.* 101. *t.* 100.

This modeft little plant is a native of Spain, Portugal, and the neighbourhood of Algiers, growing on the dry fandy hills; we have received fpecimens alfo from the rock of Gibraltar, gathered by our friend Mr. WEBER, Surgeon to a German regiment in his Majefty's fervice.

BROTERO has another fpecies or rather variety very fimilar

to

to Leucojum autumnale, which he calls *trichophyllum*, in this the petals are acute, not tridentate.

Every author who has defcribed this plant agrees that it has ufually two, fometimes one, but very rarely three flowers from the fame fpathe, yet all continue to fay fpatha *multiflora* ; it appears to be much more conftant to the character of *two-flowered* than L. *vernum* does to that of *one-flowered.*

Our drawing was taken from a fpecimen communicated by Meffrs. NAPIER and CHANDLER, Vauxhall. It flowers, as the name denotes, in the autumn, coming up without leaves, which with us feldom appear till the flowering is entirely over, fometimes not till the fpring; but in moft of the fpecimens we received from Gibraltar the leaves appear with the flower, though much fhorter than they afterwards grow; much the fame as in CLUSIUS's figure, from which all the other figures of the older authors were copied except our PARKINSON, who however coarfe, is ufually original.

Syd Edwards del. Pub by T Curtis, S'Geo Crescent Oct 1 1806   F Sanfom sculp

HÆMANTHUS MULTIFLORUS. MANY-FLOWERED BLOOD-FLOWER.

Clafs and Order.

HEXANDRIA MONOGYNIA.

Generic Charaĉter.

Involucrum polyphyllum, multiflorum. *Cor.* 6-partita fupera. *Bacca* 3-locularis.

Specific Charaĉter and Synonyms.

HÆMANTHUS *multiflorus ;* foliis tribus ovato-lanceolatis acuminatis carinatis undulatis ereĉtis, um-bella congefta globofa, petalis patentibus. *Martyn & Nodder, Monog. Ic. opt. Willd. Sp. Pl. 2. p. 25. Bot. Repof. 318. Mart. Mill. Diĉt. a. 8.*

SATYRIUM e Guinea. *Vallet Hort. t. 33. De Bry Floril. t. 44. Swert. Floril. 1. p. 63. f. 3. Morif. Hifl. 3. p. 491. § 12. t. 12. f. 11. Rudb. Elyf. 2. p. 110. f. 3.*

HYACINTHO affinis africana, caule maculato. *Seb. Muf. 1. p. 20. t. 12. f. 1, 2, 3.*

From the firft eftablifhment of a colony at Sierra-Leone, the bulbs of this beautiful flower have frequently been imported from thence, and is rather common in our ftoves.

The fpathe generally divides into three parts and is patent or refleĉted, not ereĉt, as in HÆMANTHUS *coccineus,* from which fpecies it differs alfo in radication and foliation, the fibres growing from the fummit of the bulb and the leaves, embracing

one

one another at their bafes, fo as to form a fpotted ftalk, rifing feveral inches above the ground ; in all which circumftances it agrees with Hæmanthus *puniceus.*

Being a native of fo warm a climate as the Coaft of Guinea, the bark-ftove is neceffary to its prefervation ; and even there few have been fo fuccefsful as to flower the fame plant repeatedly ; though imported bulbs will blow without the aid of artificial heat.

Introduced into the Paris garden more than two hundred years ago by M. Robin, Jun. and figured at the time by Vallet in his Jardin du Roy Henry IV. Of this inaccurate figure, thofe of De Bry, Sweertius, Rudbeck, and Morison, are more or lefs mutilated copies ; Seba's is different and better ; Nodder's is excellent, and was drawn in 1795 from a plant which flowered at Mr. Parker's, at South-Lambeth, among the firft received from Sierra-Leone ; ours was taken about the fame time, from a bulb which flowered very weakly, but on account of its fize appeared better fuited to our work. The umbel frequently contains from forty to fixty bloffoms.

Willdenow, without having feen the plant, has defcribed the peduncles to be jointed, as they are reprefented in Vallet's figure, and this would undoubtedly be an excellent diftinguifhing charaƈter, but unfortunately nothing of the kind exifts.

I Edwards del   Pub by T Curtis S Geo Crescent Oct 1 1806   F Sansom sculp

FRITILLARIA PERSICA (β). LESSER PERSIAN FRITILLARY.

****

Clafs and Order.

HEXANDRIA MONOGYNIA.

Generic Charaƈer.—Vid. N^{um}· 664.

Specific Charaƈer and Synonyms.

FRITILLARIA *perfica;* racemo nudiufculo, foliis obliquis. *Hort. Upf.* 82. *Sp. Pl.* 436. *Reich.* 2. 47. *Willd.* 2. 90. *Mart. Mill. Diƈ. a.* 2.
FRITILLARIA racemo nudo terminali. *Hort. Cliff.* 119.
LILIUM Perficum. *Bauh. Pin.* 79. *Rudb. Elyf.* 2. *p.* 183. *f.* 1. *De Bry Floril.* 63. *Swert. Floril.* 44. *f.* 1. *Dod. Pempt.* 220. *Morif. Hiƒ.* 2. *f.* 4. *t.* 19. *f.* 1. *Park. Parad.* 29. *f.* 2. *Ger. Em.* 201. *Raii Hiƒ.* 1106. *Bauh. Hiƒ.* 2. *p.* 699. *f.* 2.
LILIUM fufianum. *Cluf. Hiƒ.* 1. *p.* 130. *Hiƒp.* 130, 131.
(β) FRITILLARIA *racemofa. Mill. Diƈ.*
FRITILLARIA minima. *Swert. Floril.* 7. *f.* 2.
FRITILLARIA ramofa, five Lilium Perficum minus. *Morif. Blef.* 266.

Varies in ftature from fix inches to three feet, bearing from twelve to fifty flowers, growing in a pyramidal form.

Probably of Perfian origin; but, as we are told by PARKINSON, was introduced to this country from Turkey, by merchants trading to that country, and " in efpecial by the " procurement of Mr. NICHOLAS LETE, a lover of all fair " flowers."

The

The root is obferved to be free from the offenfive fmell of its congener the Crown-Imperial ; but to make up for this, the tafte of it is, according to JOHN BAUHIN, horribly bitter (peramarus horribilis). It appears to be perfe&ly hardy, and eafily propagated by its bulbs ; yet is lefs common than it deferves, being a very defirable flower. Bloffoms in April and May.

Our drawing was taken at Mr. WILLIAMS's, Turnham-Green.

●

N.º 963

Pub. by T. Curtis St.

VERATRUM NIGRUM. DARK-FLOWERED VERATRUM.

****

Claſs and Order.

POLYGAMIA MONŒCIA, *ſeu* HEXANDRIA MONOGYNIA.

Generic Charaɛ̃er.

HERMAPHROD. *Cal.* o. *Cor.* 6-petala. *Stam.* 6. *Piſt.* 3. *Capſ.* 3. polyſpermæ.

MASC. *Cal.* o. *Cor.* 6-petala. *Stam.* 6. *Piſt.* rudimentum.

Specific Charaɛ̃er and Synonyms.

VERATRUM *nigrum ;* racemo compoſito, corollis patentiſ. ſimis. *Sp. Pl.* 1479. *Reich.* 4. 297. *Hort. Kew.* 3. *p.* 422. *Scop. Carn. n.* 1234. *Jacq. Auſtr.* 4. *p.* 18. *t.* 336. *Mart. Mill. Diɛ̃. a.* 3. *Kniph. Cent.* 4. *n.* 91.

VERATRUM flore atrorubente. *Tourn. Inſt.* 273.

HELLEBORUS albus flore atrorubente. *Bauh. Pin.* 186. *Moriſ. Hiſt.* 3. *p.* 485. *ſ.* 12. *t.* 4. *f.* 2. *Beſl. Hort. Eyſt. Pl. Æſt. Ord.* 8. *t.* 9.

HELLEBORUS albus præcox atrorubente flore. *Park. Theat.* 216. *n.* 2. *t.* 217. *n.* 2. *Ger. Emac.* 440. *f.* 2.

This ſtately herbaceous plant is a native of Auſtria and perfeɛ̃ly hardy. MILLER obſerves, that it ſhould be planted in an open ſituation, as, when near to walls or hedges, it is apt to be disfigured by ſnails; from whence he infers that it muſt be leſs acrid than the White Hellebore, which is rarely touched by them. Flowers in June and July. Is an old inhabitant of our gardens, being cultivated by JOHN GERARD, IN 1596. We received our ſpecimen from Mr. SPON, Nurſeryman, at Egham.

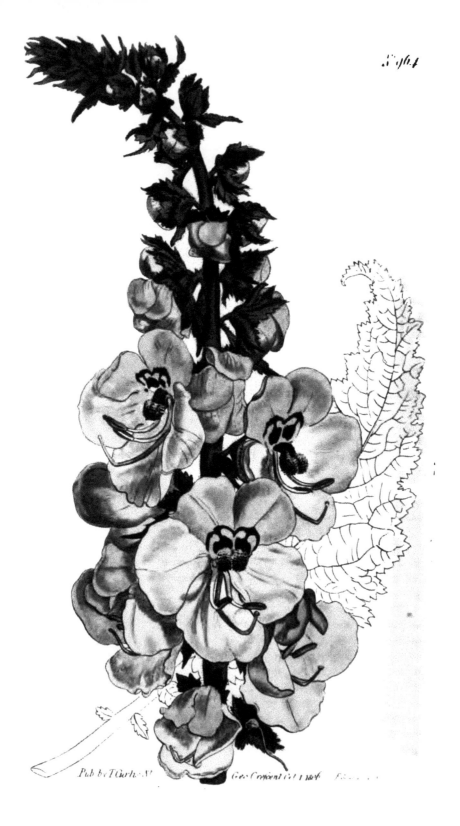

Pub. by T. Curtis, St.   Geo Crescent Oct 1 1806.   F.....

CELSIA CRETICA. GREAT-FLOWERED CELSIA.

Clafs and Order.

DIDYNAMIA ANGIOSPERMIA.

Generic Charaƈter.

Cal. 5-partitus. *Cor.* rotata. *Filamenta* barbata. *Capf.* 2-locularis..

Specific Charaƈter and Synonyms.

CELSIA *cretica* foliis inferioribus finuatis petiolatis, fuperioribus cordatis amplexicaulibus, filamentis inferioribus lævibus. *Solander MSS.*
CELSIA *cretica. Linn. Suppl.* 281. *Syƒt. Veg.* 469. *Vahl. Symb.* 3. *p.* 80. *Willd. Sp. Pl,* 3. *p.* 280. *Reich.* 3. 151. *Hort. Kew.* 2. *p.* 344. *Desfont. Atl.* 2. *p.* 57. *Mart. Mill. Diƈt.*

The figure in MILLER'S Icones (Pl. 273) generally quoted as a fynonym, does not appear to belong to this plant; fo that we do not know that any reprefentation of it has been before given.
The fyftematic arrangement of this plant has hitherto been at variance with natural affinity; if more attention had been paid to the latter, it would hardly have been diftinguifhed from VERBASCUM, feveral fpecies of which have the hairy filaments of unequal length and arranged in a fimilar manner. In one, of which we intend foon to give a figure, the upper ftamens exaƈtly refemble thofe of CELSIA *cretica*, but are three in number inftead of two: fo that here are two plants minutely correfponding, both in habit and fruƈtification, excepting that one of them has only four ftamens the other five, from which circumftance one is joined to Celfia and arranged in the
fourteenth

fourteenth clafs, the other is a Verbafcum and belongs to the fifth! GÆRTNER, who examined CELSIA *orientalis*, diftinguifhed this genus from Verbafcum by the different ftructure of the capfule, obferving that in the former the diffepiment between the cells is fingle and contrary to the valves, in the latter double, formed by the inflected margins of the valves. In this refpect too CELSIA *cretica* belongs to the genus Verbafcum; and this circumftance would have overcome our unwillingnefs to make any change in eftablifhed names, fatisfied that in uniting this plant with Verbafcum we fhould have been clofely treading in the footfteps of nature; but the examination of CELSIA *Arcturus*, which differs from the other fpecies, and from Verbafcum, in having oppofite leaves, makes us again hefitate, for in this too we find the capfule of Verbafcum. On this account we have thought it fafeft to retain this plant in its former fituation, till both genera fhall have been more accurately examined.

The CELSIA *cretica* is a fhewy biennial, readily propagated by feeds, requiring to be protected from froft. Flowers in June, July, and Auguft. Is a native of Crete, the fields about Algiers and Tunis, and faid in Hortus Kewenfis to have been introduced into this country from the Eaft-Indies by M. THOUIN in 1776.

Our drawing was taken at Mr. SALISBURY's Botanic Garden, Brompton.

d Edward del. Pub by Laurie Mc Ges Crescent Oct 1 1806 F Sansom sculp

LODDIGESIA OXALIDIFOLIA. OXALIS-LEAVED LODDIGESIA.

Clafs and Order.

DIADELPHIA DECANDRIA.

Generic Chara&er.

Vexillum alis carinaque pluries minus!

Specific Name and Synonym.

LODDIGESIA *oxalidifolia*.
CROTALARIA *oxalidifolia*. Hortulanis.

DESC. A low branched fhrub. *Leaves* alternate, trifoliate, on long filiform petioles, in the axils of which is a pair of fubulate, minute, falling ftipules: *leaflets* obcordate, mucronulate, quite entire, fmooth. *Flowers* terminal, from three to eight, in an umbel, on fhort peduncles, drooping. *Bra&es* two, minute, fubulate. *Calyx* coloured, hollowed at the bafe, fomewhat inflated, 5-toothed: teeth acute, three lowermoft rather longeft. *Vexillum,* or *ftandard,* very minute, proje&ing but little beyond the calyx, white: *Wings* about three times longer, oblong, widening upwards, obtufe, fpreading, white. *Keel* nearly equal in length to the wings, fomewhat wider, rather fquare-pointed, gaping underneath, dark purple. *Filaments* all conne&ed in a fheath which fplits at the upper part. *Ovary* oblong, compreffed, containing from two to four ovula; *Style* going off at a right angle: *Stigma* pointed. The genus may be placed in the fyftem between Genifta and Cytifus, which ought to ftand near together.

We believe that this delicate little fhrub was firft introduced into this country by GEORGE HIBBERT, Efq. of Clapham-Common, in whofe confervatory our drawing was taken.

We

We likewife received it from Mr. Loddiges, Nurferyman, at Hackney, who raifed it fome years ago from feeds he received from the Cape of Good Hope. This excellent cultivator, from his extenfive correfpondence with feveral far-diftant countries, has been the means of introducing many rare exotics into our gardens, and to his experience and fkill in horticulture, the prefervation and propagation of more, that would have been otherwife loft, is to be entirely attributed. Of his liberality in communicating his poffeffions, for the pro- motion of fcience, the numbers of our magazine bear ample teftimony, and in return, we confider it as a duty impofed upon us, thus to record his merits, by naming a genus after him. That the one we have chofen is very diftinct, we ap- prehend the fingular form of the corolla will fufficiently decide, although we have not yet been fo fortunate as to meet with a feed-veffel.

It is a tolerably hardy greenhoufe fhrub, flowers freely, and is readily propagated by cuttings. Bloffoms in May and June.

Linnæus fometimes amufed himfelf with fancying a refemblance between the genus and the perfon to whofe honour it is dedicated; and fuch conceits may at leaft ferve to affift the memory. So in Loddigefia, the minute white ftandard may be confidered as the emblem of the modeft pre- tenfions of this venerable cultivator; the broad keel, of his real ufefulnefs to fcience; and the far-extended wings, as that of his two fons,

Sic præstent virtute patri, sic frugibus ambo.

By I. Edward del. Pub by T Curtis St Geo Crescent. Oct 1.1806. F Sansom sculp.

ERICA ELEGANS. ELEGANT HEATH.

✱✚✱✚✱✚✱✚✱✚✱✚✱✚✱✚✱✚✱✚✱

Class and Order.

OCTANDRIA MONOGYNIA.

Generic Character.

Cal. 4-phyllus. *Cor.* 4-fida. *Filam.* receptaculo inferta. *An-thera* 2-fidae. *Capf.* 4-locularis. *Diffepimenta* valvulis contraria.

Obs. Foliola calycis, laciniae corollae, loculamenta, valvulaeque numero interdum duplicantur.

Specific Character and Synonym.

ERICA *elegans;* antheris criftatis inclufis, foliis fexfariis glaucis, umbellis congeftis terminalibus involucrato-braɛleatis, corollis urceolatis.

ERICA *elegans. And. Heaths.*

Descr. A low fhrub, with fhort branches growing in every direɛlion. *Leaves* ternate, but by thofe of one whorl being placed direɛlly between thofe of the next, the whole is neatly arranged in fix diɛtinɛl rows, glaucous, flefhy, acerofe, channelled underneath. *Flowers* terminal in a compaɛt umbel. *Involucre* of fix ovate, acuminate, leaves. Peduncles fcarcely as long as the involucre, with three or four large braɛtes fimilar to the involucre, deciduous. *Calyx* four-leaved, leaflets orbicular, acuminate, fomewhat fpreading, nearly equalling the *Corolla,* which is globular at the bottom, contraɛted upwards ; mouth 4-fid, fmall. · *Neɛtary* a glandular beaded circle within the ftamens. *Stamens* included ; *filaments* dilated, at both ends incurved : Anthers oblong, acute, criftate : criftae nearly orbicular minutely notched, and in this fpecies (perhaps in others) are evidently proceffes of the filaments, and no part of the anthers. *Germen* globofely four-lobed : *ftyle* ereɛt : *ftigma* capitate, included.

The

The involucre, bra&es, calyx, and corolla are all of a rofe-colour, deepeft where moft expofed to the light, the firft and laft tipped with green. A faccharine juice is fecreted in fo large quantities as to drop from the flowers.

We conclude that this fpecies is not contained in Mr. Salis-bury's monograph on this genus, in the Tranfa&ions of the Linnean Society, as it is certainly not to be found in the neigh-bourhood of *glauca*, its near affinity with which could not have paffed unnoticed. But without this clue, even if prefent, we might perhaps have overlooked it; for in fo extenfive a genus, in which the fpecies are, for the moft part, given under new names, and not arranged under different fe&ions, it is not always eafy to determine, whether a required fpecies be there or not.

Is more eafily propagated and a much freer blower than *glauca*. Our drawing was taken from a fine fhrub at Mr. Buchanan's, Nurferyman, at Camberwell, who appears to be very induf-trious in colle&ing rare plants, and obligingly communicative to fcientific inquirers.

Lightning Source UK Ltd.
Milton Keynes UK
UKHW020745251118
332796UK00002B/160/P